A Bunch of Plumbers

Many individuals in the space sciences and engineering community were surprised when the NASA administrator named the Langley Research Center as the project management center for the critical Lunar Orbiter Project that preceded the Apollo manned missions. They did not think Langley had the capability to manage the project.

Dr. Harold C. Urey, a Nobel Laureate in chemistry from the University of California at San Diego, was one of the dissenters. Dr. Urey felt this so strongly that he wrote a letter to the NASA Administrator, James Webb, asking him, "How in the world could the Langley Research Center, which is nothing more than a bunch of plumbers, manage this scientific program to the Moon?"[1]

The plumbers did very well, as you will see.

John Newcomb

HighTide
Publications, Inc.

DELTAVILLE, VIRGINIA

High Tide Publications, Inc.
1000 Bland Point Road
Deltaville, Virginia 23043
www.HighTidePublications.com

Quantity sales. Special discounts are available on quantity purchases by corporations, associations, and others. For details, contact the "Special Sales Department" at the address above.

A Bunch of Plumbers/John Newcomb, 1st ed.(revised)

Printed in the United States of America
The numerous NASA images shown in this book are in the public domain and not included in the above copyright.

The NASA numbers apply to the year and the sequence in which the negatives were archived and not the year or sequence in which they were taken.

Printed in the United States of America
Newcomb, John
A Bunch of Plumbers; How one boy's dream led him into one of the most successful space missions of the 20th century

ISBN 978-0692455852

TEC002000 - Technology and Engineering/Astronautics and Astronautics.- 1.History: Space Exploration. 2. Local History – Virginia (NASA Langley)

I would like to dedicate this book to my wife, Peggy Newcomb, whom I have left standing in the doorway of our house many times in the pre dawn light as I headed off to catch another plane.

We got married during the time I was working on Lunar Orbiter. When it was time to celebrate our first anniversary, our friends gave Peggy a one year anniversary party and me a one half year anniversary party. They thought that was all I deserved, since I had been home only one half of the time.

Table of Contents

Acknowledgments   i

The End And The Beginning
1-The End   1
2-The Beginning   3

Wallops Island And Buglia's Boys
3-Rockets And Things   9
4-Breaking Up Is Hard To Do   15
5-A Remote Outpost   19
6-The Shotput Project   25
7-Ash In Your Hash   28
8-The New Wallops   30
9-Buglia's Boys   33

Lunar Orbiter Days
10-Nasa And The Space Race   38
11-Lunar Orbiter   44
12-Who Will Get The Job?   47
13-The Rest Of The Story   53
14-Some Issues Never Die   55
15-The Lunar Orbiter Team   56
16-The Boeing Design   59
17- The Camera   62
18-What Type Of Mission Was That?   65
19-The First Of The Lunar Orbiter Missions   68
20-Bouncing Around   72
21-The Data Analyst   74
22-Roll, Pitch, Bang   75
23-Mission Design   77
24-The Newcomb Maneuver   82
25-The Forgotten "Picture Of The Century"   87
26-Everything Old Is New Again   96
27-Kentucky Windage   99
28-I Would Like To Introduce My Head Of Mission Operations   101
29-The Daily News   103
30-Moon And The Astronomers   105
31-The Debrief   107

Table of Contents

32-LO 2   109
33-LO 3   111
34-LO 4   113
35-LO 5   115
36-LO Missions   117
37-Lunar Orbiter Retrospective   119
38-Lee Scherer's Coat   121
39-Apollo Landings – One Giant Leap For Mankind   123

The Viking Days
40-The Viking Team   129
41-Joining The Team   133
42-And The Winner Isthe Martin Company   136
43-The Sol   139
44-Congressional Holdup   140
45-Landing Site Selection   148
46-Go To The Water   151
47-Mission Operations Design   154
48-The Orbiter And Lander Come To Life   159
49-Stuff It In A Box   161
50-We Are Coming In Fast   163
51-The Final Thrust   165
52-Something Has To Control This Thing   167
53-Let's Cook It   169
54-Test As You Fly   172
55-Playing With Firecrackers   175
56-A Chance Encounter   181
57-Where Do We Put It?   182
58-Getting Ready For The Final Test   193
59-The Final Test   198
We Were On Mars!   202
60-Viking On The Surface   203
61-The Site We Rejected   207
62-This Thing Ain't Working   209
63-Where Do We Park Viking Lander-2?   214
64-Let's Do It Again   219
We Had Now Done It Twice.   220
65-How Viking–1 Opened The National Air And Space Museum   224

Table of Contents

66-Lander Surface Experiments 227

67-Is There Life On Mars? 229

68-The Nasa Mars Conference 235

69-The Remaining Question Of Life On Mars 238

70-The Mission Beyond The Mission 240

71-A Bit Of Nostalgia 243

72-Viking Epilogue 246

Addendum

Life After Viking – The Unsung Heroes 247

References 259

End Notes 262

Appendices 269

About the Author 284

Acknowledgments

Even though I lived the experiences chronicled in this book, I must acknowledge the great histories about early Wallops days as well as the Lunar Orbiter and Viking projects. These include *Spaceflight Revolution* by James Hansen, *Destination Moon* by Bruce Byers, and *On Mars, Exploration of the Red Planet* by Edward and Linda Ezell. These histories are referenced many times to reinforce my memory of events. I also want to acknowledge the book, *The Vikings of '76* by Hans-Peter Biemann, which captured in pictures the story of the Viking site selection and early Viking discoveries.

I would also like to recognize my wife, Peggy Newcomb; our friend, Sharon Dorsey; and our daughter, Helen VanOrden, who read the entire manuscript, and offered comments and ideas that made the book more reader-friendly.

Acknowledgments would not be complete without recognizing Rachel Tillman, Founder and Director of *The Viking Mars Missions Education and Preservation Project*, who has given much encouragement and support.

Credit must be given to individuals from Wallops Island, Viking and Lunar Orbiter, who provided information, artifacts, and comments on various parts of the manuscript.

These include Dale Shellhorn, Gentry Lee, Woody Lovelace, Burt Lightner, John Graham, Bob Buchan, Paul Seamers, Fil Cuddihy, Matt Grogan, Otis Childress, Bill Michael, Bill Boyer, Gary Price, Jim Hall, Ed Guinness, Dempsey Bruton, Barry

DiGregorio, Joe Gowdey, Bill Jones, Len Clark, Dr. Gilbert Levin and Bob Tolson. I offer a special thanks to Teresa Hornbuckle, the NASA archivist who researched the NASA pictures and provided much help and support.

I also offer a special thanks to Norman Crabill for reviewing the entire manuscript and providing very useful thoughts and comments.

1-The End

This date is forever emblazoned in my memory. It was July 20th, 1976, 5:12 a.m. Pacific Daylight Time.[1] We were in the Space Flight Operations Facility (SFOF) at the Jet Propulsion Laboratory in Pasadena, California.

The place was a madhouse. Tears were welling up in my eyes. I was almost incapable of moving for a few minutes. I also had to remember to breathe. Everyone was smiling, yelling, hugging, and shaking hands.

Noel Hinners, the NASA Associate Administrator for Space Science, would later say at the press conference, "I had tears in my eyes this morning for the first time, I guess, since I got married. It was really an emotional experience."[2]

Gentry Lee, leader of the operational mission design effort, had been interpreting the lander trajectory parameters for the science team. He was, for a moment, speechless.

Tom Young, who was head of Mission Operations, was laughing and cheering, all at the same time. Jim Martin, the Viking Project Manager, was a mountain of a man at six feet four inches, with piercing blue eyes and a white crew cut. He had led us through the eight years to arrive at this point. He had the biggest smile on his face that I have ever seen.

The Viking lander team lead, Rex Sjostrom, had just announced that Viking 1 had landed safely on the surface of Mars nineteen minutes earlier. (It took the radio signal nineteen minutes traveling at the speed of light to reach Earth.) This marked the first time anyone from this Earth had successfully landed a vehicle on the surface of Mars.

This project was the culmination and release of eight years of stress, hard work, many night and weekend hours, and many personal sacrifices...all accomplished by an incredible team of scientists, engineers and support personnel who now uniformly exhaled a loud sigh of relief.

And I was privileged to be there and to have been part of this monumental project.

Note: The Russians landed a vehicle on Mars in 1971, but it survived for only twenty seconds and provided no useful data.

2-The Beginning

This is the story of a young fellow from a rural community who had a fascination for airplanes and anything that flew. In grammar and high school, I filled the house with model airplanes that I was constantly building and crashing. I read aircraft training manuals like other kids read comic books.

My mother was overjoyed when our senior class went to Washington, D.C. and visited the Smithsonian. She thought I would come back talking about something other than airplanes. But it did not work. The only thing I could talk about when I came back was Lindbergh's "Spirit of Saint Louis".

When I learned that a NACA (National Advisory Committee for Aeronautics) laboratory (Langley Research Center) was not far from my home, I decided I would work there. At the tender age of fifteen, my future was charted. I would work for NACA at the Langley Research Center in Hampton, Virginia.

The reasons for that decision matured as I did. However, my decision never wavered. So it was that I became passionately interested in science and engineering and decided to go to Virginia Polytechnic Institute. There I majored in Engineering Mechanics (now called Engineering Science and Mechanics). I then entered the Cooperative Engineering Program, working for a quarter and attending classes for a quarter.

By my senior year, I was promised a permanent job at the

Langley Research Center in Hampton, VA. This center was now part of the newly formed National Aeronautics and Space Administration (NASA).

This offer sealed my future. I was so determined that NASA Langley was the only place I wanted to work that I did not even interview with other corporations in my senior year at Tech.

In the early'60s, the aerospace industry was growing rapidly in response to the launch of Sputnik (the first earth orbiting satellite) in 1957 by the Russians. By 1962, the year I graduated from VPI, the Russians had successfully launched two spacecraft toward the Moon. In 1959, the Russian Luna 2 impacted the lunar surface. Also, in 1959 the Russian Luna 3 took the first pictures of the Moon's far side.

Due to the aerospace industry growth, the engineering college campuses were crawling with corporate recruiters enticing seniors to sign up. My reaction to their invitations was, "Interviewing me is a waste of your time and mine. I know where I am going, and there is nothing you can say to change that."

And so I came to NASA Langley Research Center as an engineer in 1962. There I had the privilege of working on two projects that marked firsts in our nation's, and the world's, history.

The first was the Lunar Orbiter Project, which placed five spacecraft around the Moon between mid 1966 and 1967. The project provided critical information to support the landings of the Apollo vehicles on the lunar surface. The second was the Viking Project, which successfully landed the first spacecraft on the surface of Mars in 1976.

My mother always told my wife, Peggy, that what I did was a mystery to her. I only wish that she were alive to read this book. Maybe it would give her some understanding.

3-Rockets And Things

This is heaven. Here I am, a college student on work assignment at Langley as part of the Cooperative Engineering Program. I am where I want to be, and I can taste the excitement. I am about to take my first trip to Wallops Island on the northern end of Virginia's Eastern Shore.

This is Langley's rocket launching site. While I am here, I will help prepare and launch various payloads (experimental spacecraft) to be launched on the front of the rockets.

It is a beautiful fall day. The breeze is blowing. Salt spray is coming over the side of the boat as we travel through the marshes to the Wallops Island Launch Site.

I had flown in from Langley Research Center and landed at the Chincoteague Naval Air Station on the mainland, taken the jeep to the dock, and boarded the LCM (a Navy Landing craft) for the trip to the Island. Now the sound is deafening as we roar through the marshes in the LCM, as if to storm the beaches at Iwo Jima.

During the trip from Langley, I was the courier for the payload we were going to launch. It was wrapped carefully in layers of foam, placed in a waterproof case, and handed over to me for safe transport to the launch site. I was seated beside it as we flew in the C 47 to the air station. I made sure that it was buckled in safely for the flight.

Now I am carefully carrying it during the boat journey. When we reach the Wallops shore, the jeep and driver are waiting. I get in the jeep and head for the payload processing hanger, where I safely deliver my cargo.

My "Yacht" for the first trip to Wallops
Credit: Joseph Shortal

During the next several days, I had my first experiences preparing for launch. While I was waiting for our payload to be prepared, I saw that another four stage rocket was being readied. Since I did not have anything to do at that point, I asked if I could help.

The answer was, "Certainly. We would like you to help bolt together the solid fuel rocket stages while the technicians make the final tests on the payload to ensure it is ready for launch." Preparing for launch in those days was quite a different process that you would see today.

There were some rules, like connecting a grounding strap whenever you entered a payload or rocket processing area, but there were not many. Whoever was available and could do the next job did it.

The rocket being readied was a four stage rocket, an Honest

John-Nike-Nike-Recruit (where each name stands for a solid fuel rocket stage). These rocket test flights were useful, because we could get aerodynamic data from various shaped models at velocities and conditions that we could not achieve, at that time, in the wind tunnels. As these rockets and models flew downrange, they would radio back the aerodynamic data to ships, planes, or the launch site. After the flight was completed, the payload would simply fall into the ocean. These same types of rockets would also be used for testing various spacecraft and instruments.

As I bolted these stages together, a terrible rainstorm rolled in. I could not abandon the rocket, since the stages had not been fully secured to each other, so I continued to sit astride each stage and strap on the next one until I got the job done. This resulted in a well connected rocket and one of the worse colds I have ever had.

Later, when the payload was secured and we were ready for launch, I asked Derwood, the lead technician, if I could arm the rocket. Derwood said, "Sure."

I suppose he was happy to hear this request, since it was his job to arm all of the rockets for launch. Derwood had a wooden arm from the elbow down, as the result of an accident on the job. One day when he was arming a Deacon single stage rocket, the rocket decided it was time to leave and fired prematurely. The tail fin of the Deacon caught and launched Derwood's hand along with the payload and broke a bone in the other hand.[1] So Derwood was very happy to honor this request, even if it was from an inexperienced college kid who didn't know beans about what he had just asked.

Four Stage Rocket Ready for Launch
Credit: NASA 1960 L 05828

After we had gone over the process in detail, I stepped up to the arming box and took off the lid. It was at this time I looked down all four stages of the rocket I was about to activate and visualized all of the high power propellant they contained. Did I really want to do this job, especially when I realized that my head and upper torso were all lined up perfectly with some of the fins?

My line of thought was broken as Derwood started giving the instructions. I heard him say, "Okay, now you see the extra nut on pin 4. Take it off of pin 4, just as we talked about." Thus began one of the most intense conversations of my life:

John	Derwood
Okay. I got it.	Put the nut in the tray so you will have both hands free.
Done	Now loosen the nut over the red wire we talked about on pin 6. Remember to hold the red wire with the other hand so it remains in contact with the pin it is on.
I'm doing it. Now the nut is off.	Good, now drop the nut in the tray, as you won't need it anymore.
Done.	Now carefully remove the red wire from pin 6 and place it on pin 4. Remember DO NOT ALLOW the red wire to touch any other pin in the arming box. Now where are you?
I have the red wire in my hand.	Good. Steady the hand holding the red wire with the other hand before placing the wire on pin 4. We don't want any misses here. Okay, are you holding the red wire with your right hand and steadying with the left?
Yes	Are you steady?
I think I'm steady, just a little nervous.	Okay, just stay steady. Now are you ready to put the red wire on pin 4?
I think so.	Okay, just place it on the pin
Okay	Now continue to hold it on the pin with the right hand so it continues to make contact, and pick up the nut with the left hand. Place the nut on pin 4 and tighten with your fingers. Now use the wrench in the tray and tighten the nut one half turn. Remember don't let the wrench touch any other part of the box. It is insulated, but there is no use taking chances. Put the box top back on and secure it.

So went the first, and I might say, only time I armed a rocket.

After the box top was secured, a smiling and relaxed Derwood came over to congratulate a slightly unnerved nineteen year old.

By the way, that launch and payload operation went very well, and we got the aerodynamic data (lift, drag, and so on) we needed from our model, just as planned.

I have thought about this story many times since that day. I am sure Derwood disconnected all circuits, as was normal. But did he actually disconnect some additional wires? I will never know. And Derwood, sadly, is no longer with us to ask.

Also, I should mention a side note on Derwood. He was quite an accomplished violinist. The fact that he now had a wooden arm did not deter him at all. His wooden arm was equipped with a split hook on the end that would open and close. He taught himself to play the violin holding the bow with the split hook, and he continued to perform in various churches and at different gatherings in the area.

4-Breaking Up Is Hard To Do

It was now time for us to begin working on our payload, which was a flight test of a separation device for the Polaris warhead. During this time, the Navy was developing the Polaris missile. These missiles, containing nuclear warheads, would be carried and launched from beneath the surface of the ocean by submarines.

During the development of the Polaris in the mid 1950s, test launches revealed the missile had a problem. Everything would go along fine for a short period of time after the Polaris warhead was released from the final rocket stage. Then the last rocket stage would start bumping the warhead in the tail. This was due to the fact that the Polaris had a blunt nose that produced a lot of drag. After separation, the last stage would ride in the wake of the Polaris. In a way, the last stage was drafting behind the payload just as a racecar driver drives right behind one of his team members in order to save his engine for the final laps of the race.

The separation mechanism used to detach the warhead from the last propellant stage was standard for the time. This consisted of a ring type clamp (called a Marman clamp) that held together the flanges of the warhead and the last stage.

This clamp was tightened over the flanges by the use of two explosive bolts that pulled the two halves of the clamp together. When separation occurred, a signal was sent to the two bolts causing them to explode, releasing the clamp. A piston, driven by compressed gas, then pushed the two stages apart. It was low tech, and in most cases it worked.

Only it did not work with the Polaris.

NASA Langley was asked to solve the problem, and the job was given to an engineer named Jim Hall. They gave the problem to the right guy.

Jim was a robust character who had volunteered and flown for the Royal Canadian Air Force (RCAF) in World War II before the US got into the war. He had volunteered for the RCAF because he wanted to fly Spitfires (high performance British fighter aircraft). However, he never got the chance to fly the Spitfires, because he spent the whole time flying multiple engine aircraft, mostly bombers, over Europe, Africa, and Asia. He also flew Wellingtons (British bombers) at night fifty feet over the water, dropping depth charges on German submarines.

Jim was one of those project engineer/managers to whom you gave a problem and then got out of his way. He was a no nonsense kind of guy. He was also noted for his eloquent, though politically incorrect, vocabulary. He had a bit of a temper, which often gave rise to the use of this jargon.

In one meeting Jim got irritated, leaned over the table, looked the opposition straight in the eye, and said, "Now look here, you spherical son of a bitch." The fellow was a little taken back.

"I understand son of a bitch, but why spherical?" his opponent asked.

"Because you are perfectly symmetrical. Viewed from any angle or direction, you are a son of a bitch!"

Jim, working with John Graham, solved the Polaris problem very simply. They had the designers place two small rockets at the front of the last rocket stage of the Polaris. These small rockets were pointed at right angles to the long axis of the main rocket. After the bolts fired, the clamp that held the last stages and the warhead together came off, the piston pushed them apart, the small rockets fired, and the last rocket stage would be forced at right angles to the flight path and would fly/tumble away from the Polaris.

We have used this concept many times since that test, but to my knowledge, this was its inaugural use. The payload that I had carried to Wallops was the dummy model of the Polaris warhead that we used to test Jim's design.

Polaris Warhead and Nike Ready for Launch
I am standing at the bottom of our rocket.
Probably not a good place to stand.
Credit: NASA

5-A Remote Outpost

During World War II, everyone was developing missiles. By the end of the war, they were being used by the English, Russians, United States, and the Germans. Starting in the early 1940s, the Navy, Army, and the Air Force requested NACA to develop a missile test site. After much study, it was decided that Langley would develop the site. Langley originally considered locating it on the Marine base at Cherry Point, North Carolina. However, during the time these studies were being performed, it was realized that the Navy was planning to condemn and purchase a small island just to the southeast of the Chincoteague Naval Auxiliary Air Station on the upper portion of Virginia's Eastern Shore just south of the Maryland state line.

It was called Wallops Island.

In early 1945, a delegation from Langley and NACA Headquarters was dispatched to examine the possibility of developing a launch site at either Wallops Island or Cherry Point, North Carolina. In both cases, they would need to use some of the existing facilities and build additional ones on the sites. The delegation found that launches could be made over a wider range of directions from Wallops than Cherry Point. That would make Wallops a superior launch site. They also found a warm reception at the Chincoteague Naval Auxiliary Air Station, in contrast to the reception at Cherry Point. Joseph Shortall relates in his History of Wallops Station, "Reception of this group by the Marines at Cherry Point was very cool. In fact, the operations officer said that he would file an official protest if Langley planned to locate there."

So later in 1945, the development of Wallops Station, (soon to be called The Pilotless Aircraft Research Station) started. The Pilotless Aircraft Research Division was started at Langley Research Center to provide the guidance for the development of the Station and to develop the test vehicles that would be flown from the station.

In 1958, I came to Wallops (with the model Polaris warhead) for the first time. It was just after my freshman year at Virginia Polytechnic Institute (VPI). Those early days on Wallops were thrilling for me.

Wallops still looked like a remote outpost. This feeling of being on an outpost added to the excitement that goes along with launching rockets. The island had only the bare essentials. It had the launch pads, the bunkhouse, the launch control center, the missile storage bunkers, and the vehicle test and assembly buildings. I enjoyed every bit of it, from just watching to participating in all of the aspects of the launch operation.

One of the best shows in town was Roy, the operator of the skin tracking radar. During launches, we had radar signals coming back from the payload, but we also used a skin tracking radar that bounced signals off the skin, or covering, of the payload. This gave us an independent measure of the location and velocity of the payload.

We used a World War II radar that was located behind a blast fence near the launch site. The radar had a gun sight mounted on it, and the radar operator had to look through it to keep the payload in the sight and the radar viewing the vehicle as long as possible. The operator only had a second or two to lock the radar on the vehicle and initiate the auto track.

The operator had to use his hands and feet to manually point the radar. Using foot pedals (like the pedals of a bike), he would move the radar in azimuth and use wheels with handles on them to move the radar in elevation.

This might not sound like such a challenging endeavor until you realized that these solid rockets went from zero to approximately 1000 miles per hour in 3 seconds.[1] This is approximately 15 g's, or fifteen times the acceleration due to gravity.

Watching a solid rocket launch is quite different than watching a liquid fueled rocket launch. Instead of the gradual lift-off of the liquid-fueled rocket from the pad, the solid rocket launch is more like watching a bullet leave the launch site.

Watching one of these launches for the first time with Roy as the radar operator was an amazing experience:

5 Minutes to Launch: Roy squirms around in his seat and jerks the azimuth pedals and elevation wheels first one way and then the other.

1 Minute to Launch: Roy sights the radar on the vehicle sitting on the pad, and quickly traces a small arc in the direction the rocket would go.

30 Seconds to Launch: Roy sights the radar a little ahead of the vehicle, and nervously waits for the launch. His hands opening and closing around the handles of the elevation wheels, his feet shuffling back and forth on the azimuth pedals.

It was then that I realized the importance of all of these maneuvers. Once the vehicle was launched, you could not see his hands or his feet. They were just a blur while he maneuvered the radar to keep it on target with the rocket. In fact, Roy was the only radar operator who could track some of the faster rocket launches.

The stories told at Wallops were non stop, especially in the bunkhouse late in the evening as the poker games started. Some would tell about the time the payload separated from the last stage just prior to launch.

At that time everyone was in the launch control bunker and could not hear a thing that was going on outside. Also they could not see the launch platform from there.

The Countdown Went Something Like This:

- Coming up on launch: final countdown

- "10, 9, 8, 7, 6, 5, 4, 3, Bang'"

- The separation device fires and the payload falls on the launch pad.

- "2"

- From flight controller: "Give me the accelerometer reading! Wait! That doesn't make sense!"

- "1"

- "Fire!"

- Rocket takes off.

- Flight controller: "Oh shit! The reading looks crazy."

- Question to range safety: "Do we destruct? Skin tracking says we are on flight path."

Finally, after much discussion and expletives, Roy (the skin tracking radar operator) realized they had sent the propellant stages on a beautiful ride downrange and still had the payload sitting on the launch pad.

Another time, a launch was going well. The rocket was heading downrange, as it should be.

Now The Reporting Went Something Like:

- The telemetry operator: "range 14 miles, range 22 miles, 38 miles" and then something went wrong.

- Then you heard "range 39 miles, range 38 miles"

- It was going straight up and over: "range 32 miles, range 25 miles, range 12 miles, range 1 mile."

Well, the rocket flew straight back over the launch control blockhouse and landed in an intersection in the town of Chincoteague.

On another launch, there were three rockets ready for flight but still in the horizontal position. The technicians wanted to go back on the pad and check some wiring. The range Safety Officer had just informed Dempsey Bruton (the Range Manager) that a lightning storm ten miles away was coming in from the west. Dempsey told the technicians in no uncertain terms to stay off the launch site.

Just as the technicians were grumbling about Dempsey's announcement, their complaining was interrupted by the horizontal, unplanned launch of the three rockets as a burst of lightning sent charges down all three of the firing leads.[2]

Then there was the time we dropped a payload in the Boston Harbor. These mishaps were just part of the early trials of the rocket program, but they did not make the locals very happy.

6-The Shotput Project

Shotput Ready for Launch
Credit: NASA 1960 L 00485

For all of the misadventures at Wallops, we were maturing and learning how to launch rockets, how to build spacecraft, and how to carry out experiments in space. Wallops Island played a major role in those early space efforts. One of the early projects was in support of the Echo project, and the supporting project was called Shotput.

A Langley researcher, William (Bill) O'Sullivan, proposed the idea of launching a large balloon and tracking it to determine the density of the upper atmosphere. We needed to know the density so we could determine how rapidly satellites would spiral in toward the Earth due to the drag caused by the friction of the molecules in the atmosphere.

With the launch of Sputnik in October 1957, it became very important to demonstrate that the US could launch a satellite. Now these two factors came together. Project Echo was formed to launch a 100 foot diameter balloon into Earth orbit and track it to determine the atmospheric density, starting at 1000 miles altitude and continuing to track the balloon as its orbit decayed. The plan was also to bounce radio signals off the balloon and perform long distance communications hence the name, Echo.

There was another interesting feature. The balloon would not only serve the science and technology requirements, but it also would be visible (due to reflected sunlight) to observers in many parts of the world as a constant reminder that the United States also had capability to go into and utilize space.

Before such a balloon could be launched there were many questions to be answered:

- How do you fold the balloon?

- How do you inflate a huge balloon in space?

- How do you package it so the packaging will get out of the way when necessary and not damage the balloon?

- What materials do you use to make the balloon?

The Shotput Project was formed with Norman (Norm) Crabill, as project manager, to provide the launch vehicle that would allow the experimenters to answer these questions. The project would provide for a test of the final Echo concept by using a two stage rocket to propel the payload up to 200 miles, at which point the canister would open and the balloon would inflate. There were to be five Shotput launches with the 100 foot balloons.

The first launch test did not go well. The balloon exploded during the inflation and showered the east coast sky with shiny bits of aluminum coated Mylar, glistening in the glow of the setting sun. This was not one of the balloon engineer's better days. Actually, Norm's rocket worked well, but the experiment failed in a much too visible way. However, as the testing proceeded and knowledge was gained, the capability required to design and deploy the balloon was soon developed.

Finally, the 100 foot diameter balloon was launched on a Thor Delta rocket from Cape Canaveral on August 12, 1960 and successfully placed in a 1,000 mile orbit that the entire world could see, compliments of the work done at Wallops.[1]

7-Ash in Your Hash

1958 Telling stories, playing poker, and walking on the beach were typical evening activities at the bunkhouse and cafeteria. In fact, the only large room was the cafeteria, so all social functions (of the kind just described except for walking on the beach) were held there.

This was truly a bunkhouse. The rooms were large enough for two sets of bunks (upper and lower), one on each side of the door. The space between the bunks was the width of the door itself. You had to notify your bunkmates if you were getting up so you wouldn't butt heads or stand on someone's shoulders.

Then there was the wake up call, performed each morning by Sam, the cook, who would throw open the door and yell something like, "Alright you pussy footing city slickers! You going to sleep all day?"

About that time, you realized that Sam had been on one of his all night fishing expeditions. You noticed that because he was covered with fish guts and scales. (Did I mention that he was the cook?) Actually, you did not have to open your eyes because the smell would let you know.

By the time we got down to the cafeteria, Sam would have miraculously cleaned up and prepared the "best breakfast on the Island." Of course, it was the only breakfast on the island, but it was good: corned beef hash; bacon; toast, etc.

The only thing you had to be a little cautious about was his cigarette. He is the only person I have ever known who could smoke a cigarette down to the very end and keep the complete ash. So here's the cook, scrambling eggs and talking to the fellows in the breakfast line, with a two inch ash curling downward and dangling off the end of the cigarette.

Of course, the big joke was: Who is going to get the ash in his hash?

8-The New Wallops

A few years later in the early sixties I returned to the Island. Things were a lot different. The space race was even hotter than before. We had now flown eight unmanned Mercury capsules from Wallops Island to test the capsule reentry and landing characteristics. This was called the Little Joe Project. Langley had flown the Mercury Capsule with its first human passenger, John Glenn, (later a U.S. Senator from Ohio) in February 1962 from Cape Canaveral.

We were now NASA: the NACA label was gone. Langley had not only managed the Little Joe development and flown it from Wallops, but it was managing the Mercury project, now being flown from the Cape.

Langley had also developed the Scout launch vehicle. This was a relatively inexpensive way to get a 100 pound spacecraft in low Earth orbit, and it was a very successful vehicle. After some early failures and a steep learning curve, the Scout would be able to boast over one hundred launches in a row without a failure. Quite a record.

The Scout even ushered in international cooperation. In the San Marco Project, Wallops launched an Italian payload on the Scout and trained the Italians on the launch operations.

The group I was in was monitoring the development of the guidance and control parameters that were fed into the Scout to allow it to deliver the various payloads into the desired orbit. Having completed this job a few weeks earlier, we now rode over to watch the launch and witness the results of our work.

The Scout, Ready for Launch
Credit: NASA 1960 L 03966

As we neared the launch site, we drove over the newly built causeway. The bunkhouse was completely renovated. The cafeteria had been upgraded and transformed into a fancy eating establishment, although still with cafeteria type food service.

Excitement was in the air! This was one of the early launches of the Scout vehicle, and a lot of reporters were on the scene to cover the launch. There was nothing we could do but wait and watch with the rest of the NASA troops and the reporters. I decided to get a bite to eat before going into the control center. I walked up to the cafeteria line, and then I heard, "What'll it be young fellow?"

There was Sam. They had cleaned him up quite a bit, and he actually looked pretty good. He was not smoking, so it looked like the game of Russian roulette, with the cigarette ash, was over. I hesitated before ordering, as I realized that the old Wallops was gone forever. I looked around for a moment, just letting the realization sink in. This was a new Wallops. We were a fancy launch site now with reporters, causeways, and TV cameras. No more LCM's. No more Jeeps. No more bunkhouses.

"You gonna eat or just stand there?"

"Yeah, Sam, let me have . . "

As he plopped the food on the plate, it was comforting to realize that one part of the old Wallops was still there, at least for a while.

"Thanks Sam, it's good to see you."

9-Buglia's Boys

Jim Buglia (pronounced "bull ya" and better known as Bull) came to Langley from Penn State in 1953. He was one of the best physical mathematicians I have ever known. I use the term "physical mathematician" because Bull was excellent at visualizing what an analysis was telling him. It wasn't just a mathematical result. To him, the result of an analysis had meaning in a very precise, physical way.

Not long after he came to Langley, he won his reputation for being an exceptional analytical problem solver. In the early Scout Launch Vehicle days, there was a problem with some of the spin dynamics associated with releasing the payloads. Bull was given the problem and sent to Ling Temco Vought (LTV), which was the contractor developing the Scout, to see if he could solve it. He stayed in the LTV plant about a month, and when he left, he had modeled the problem mathematically and had developed the solution.

After he came back to Langley, LTV called and offered him a job at double his salary. Langley said they could not match that offer, but said they would give him a raise each year as quickly as the federal regulations allowed. They would let him form his own group, hand pick his people, and work on any problems he desired. Quite an offer! Bull thought about that and decided to stay at Langley.

Thus "Buglia's Boys" was formed.

George Young, one of Buglia's Boys, would act as Bull's replacement whenever Bull had to be away. All of the guys in the group were free thinkers, not hampered by convention or normal modes of operation.

Buglia and the Boys
Left to Right in Foreground: Jim Buglia, George Young, John Unangst with Larry Hoffman and me (the short guy) in the background.
Behind the camera was Jessie Timmons.
Later we added Richard Green and George Lawrence

George was the leader in this regard. One week while Bull was away, leaving George in charge; he totally reorganized the group and sent memos to the management explaining the reorganization. Needless to say, Bull reorganized the reorganization when he returned.

George recruited me into the group during my last student work quarter at Langley, so when I came back to Langley as a graduate engineer, I became one of Buglia's Boys.

Working in this group was like being in a graduate school seminar class all of the time. The group had gotten the reputation for being good analytical problem solvers, and so other areas of Langley would bring us problems to work.

We each had various issues on which we were working. As someone got a new problem, he would come into the offices and announce, "Okay guys, get a cup of coffee; we are going to have a board talk." He would go to the blackboard and outline the issue he was trying to solve and propose the way to address it mathematically. Then someone would say, "Whoa, you can't make that assumption," or, "You just defied the first law of thermodynamics."

It was an awesome learning environment and experience. Through this group I had my first experience in large space projects of international interest.

During this time, NASA decided to sponsor and send certain individuals to a summer symposium on dynamical astronomy so we could get a big gulp of celestial mechanics and the latest in mathematical methodologies. I always used to say it was to teach us how to spell "Kepler" (a German mathematician and astronomer).

I was chosen from Langley and was sent off to Cornell University for six weeks. This was one of the most amazing academic experiences I have ever had. I had always been a student of astrophysics and had read, taken notes, and reread George Gamow's books such as *One, Two, Three, Infinity,* which is the history of mathematics, and *Until the Sun Dies,* which is about stellar evolution. Now, looking through the telescope at Cornell, I had Dirk Brouwer, who had written the books on astronomy, talking in my ear about what I was seeing. This was awesome!

I came back to Langley with several mathematical problems I wanted to work on, but that wasn't in the stars! I came through the door and proposed these to Bull.

Bull's response. "Not now, Newk. We have real problems to solve. They involve a spacecraft called Lunar Orbiter that will photograph the lunar surface to find suitable landing sites for the Apollo mission."

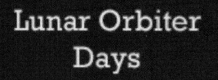

Lunar Orbiter Days

10-NASA And The Space Race

President John F. Kennedy announced his support for the Apollo program during a special address to a joint session of Congress on May 25,1961[1]:

> "I believe that this nation should commit itself to achieving the goal, before this decade is out, of landing a man on the Moon and returning him safely to Earth."

This was the public announcement. However, NACA had already been transformed into NASA in 1958 with the idea of supporting this goal. In addition, the charter of NASA expanded the old NACA charter that just talked about "conducting research and experiments in aeronautics." It did not mention space.

The NACA Charter (developed in 1915 when NACA was formed) reads in part, "to supervise and direct the scientific study of the problems of flight, with a view to their practical solution, and to determine the problems which should be experimentally attacked, and to discuss their solution and application to applicable questions."

With regard to the establishment of laboratories, the NACA charter states: "in the event of a laboratory or laboratories, either in whole or in part, being placed under the direction of the committee, the committee may conduct research and experiments in aeronautics."

NASA had now been formed, and the charter specifically introduced space exploration and experimentation. A small portion of the National Aeronautics and Space Act of July 1958 is shown below:

Functions Of The Administration

Sec. 203. (a) The Administration, in order to carry out the purpose of this Act, shall:

Plan, direct, and conduct aeronautical and space activities.

Sec. 103. As used in the Act the term "aeronautical and space activities" means

A: research into, and the solution of, problems of flight within and outside the Earth's atmosphere

---

The space race was in full swing. The Apollo program's goal to place a man on the Moon was well underway. One of the many issues facing the Apollo Program was the need to understand the nature of the lunar environment and surface:

- Was the surface hard? Soft?

- Could the Lunar Excursion Module (LEM) simply drill a hole in the surface with its rockets, as it landed, and have the lunar material fall in on it as it descended into the well it was drilling?

- Where should we look for landing sites?

- What was the average slope of the terrain in the proposed landing areas?

- Were there large rocks in these areas?

- What was the lunar gravitational field like?

- What type of radiation might the astronauts encounter as they approached, orbited, and landed on the Moon?

- What was the micrometeoroid flux in the vicinity of the Moon?

These were some of the questions that needed to be answered before the US would commit humans to this endeavor.

In order to understand the surface slopes and sizes of rocks, we needed photography down to approximately one meter resolution. Even with the best resolution we had from Earth at that time, you could have several aircraft carriers sitting on the moon and we could not have seen them.

Three different projects with three different types of spacecraft were initiated to answer these questions. The first of these types of spacecraft was known as Ranger.

The Ranger Spacecraft
The dark area toward the end away from the
solar panels is where the cameras are
located Credit: NASA Science Missions

The nine Ranger spacecraft in the series were designed to be launched on Atlas Agenas, (two stage launch vehicles) to fly straight into the lunar surface and take pictures of the planned Apollo landing sites, up until the point of impact.

All of this occurred at a time when NASA was learning how to build launch vehicles and spacecraft. In the first two Ranger missions (August and November of 1961), the second stage Agena experienced problems and failed to place the spacecraft on a translunar trajectory. In both cases, the Ranger spacecraft eventually burned up as they came back into the Earth's atmosphere.

The third, fourth, fifth, and sixth Rangers all experienced some malfunctions and did not return data. Then, in July of 1964, Ranger 7 took the first close up pictures of the Moon as it flew into and crashed in an area near Mare Nubium on the Moon's surface. (10.7°S, 339.3°E). This area was quickly renamed the Mare Cognitum, meaning "the sea that has become known."[2].

Ranger 7 returned over 4000 pictures from its six TV cameras during the last 17 minutes of flight. Rangers 8 and 9 followed successfully in February and March of 1965, landing in Mare Tranquillitatis (2.71°N and 24.8°E) and Alphonsus Crater (12.9°S and 357.6°E). [3]

After these missions, we had some close up pictures of the Moon (up to one half meter resolution).

While the Ranger missions gave the close up photographs of the Moon's surface, we still needed to land and see what the surface was like. This was the mission of the Surveyor program.

On June 2, 1966, Surveyor 1 successfully landed in the Crater Flamsteed (2.5°S, 43.2°W) and became the first US spacecraft to land softly on the lunar surface.

Surveyor
As it would look on the lunar surface
Credit: NASA Space Science Data Center

Surveyor 2 did not fare as well, had an engine failure, and crashed on the lunar surface. Surveyor 3 successfully landed in Oceanus Procellarum (2.9°S, 23.4°W) on April 20, 1967. Surveyor 4 lost radio contact 2½ minutes prior to touchdown and was never heard from again, crashing in Sinus Medii.

Surveyor 5 successfully landed in Mare Tranquillitatis (1.4°N, 23.1°E) on September 11, 1967. Surveyor 6 headed back to Sinus Medii and this time successfully landed on November 10, 1967.

Surveyor 6 actually performed a hop maneuver in order to examine the results of firing the engines close to the lunar surface.[4]

After examining the Apollo sites, the final Surveyor 7 successfully landed at a site of primary scientific interest that was the Crater Tycho (40.9°S 11.5°W) on January 10, 1968.[5]

11-Lunar Orbiter

1963 The third project to be initiated was Lunar Orbiter. The purpose of the Lunar Orbiter (LO) Project was to take pictures of the planned Apollo landing sites so maps could be made. These maps would be used to select the Apollo landing sites. The photographs would also be used to develop mosaics of the lunar surface so the astronauts could train by looking at the mosaics as they attempted to land. Since Lunar Orbiter was to be in orbit about the Moon for months at a time, it would also answer questions about the lunar gravity field, the micrometeoroid flux, and the radiation environment.

With any large proposed project, many studies are performed prior to the project's actual initiation in order to answer the major questions about the pending effort. The pre Lunar Orbiter effort was no different.

One particular study done by Space Technologies Laboratory (STL) examined how the mission could best be accomplished. They concluded that a spinning satellite would be the most cost effective way to photograph the Moon and meet the needs of Apollo. This study, which favored the spinning satellite design, would cause much heated discussion over the next several months.[1]

In the spinning satellite concept, the spacecraft is oriented in a certain direction and then it is put in a spin with the use of small rockets. It then maintains its orientation in space indefinitely, acting like a gyroscope.

The second approach is to have the spacecraft three axis stabilized. In order to do this, the spacecraft has to have two sensors and an active control system. The first is a sun sensor that can recognize the sun, and the second is a star sensor that can recognize some particular star like Canopus, which is very bright and located almost 90 degrees down from the plane of the rotation of the Earth around the sun. The control system would consist of control rockets to orient the spacecraft, and gyroscopes to measure the rotations about the spacecraft axes. With a three axis stabilized spacecraft in proper orientation, you know that the spacecraft points toward the sun and has rolled to a given orientation. It is possible to move to another orientation by commanding rotations about any of the three axes. In addition, the sensors will tell the spacecraft when it begins to drift off the proper orientation, and the small control jets will fire to bring the vehicle back to the desired alignment.

While the mission studies were under way, NASA Headquarters in Washington was holding discussions to determine what center should have the responsibility for the Lunar Orbiter Project. There was a consensus building that the Jet Propulsion Laboratory had all it could handle with the Ranger and Surveyor missions. Headquarters had been pushing Langley to take the Lunar Orbiter job.

Meanwhile, Langley had been performing its own studies to determine if they wanted to take on the responsibility of the Lunar Orbiter project. There was a lot at stake, including not only the Lunar Orbiter Mission but also Langley's reputation. Langley had never managed a mission such as this. What if we failed? Did we have the manpower and needed skills to do the job? Could we get support from the research divisions to support the Lunar Orbiter Project?

These studies and concerns led to many discussions and

meetings. Finally, in one meeting with Center Director, Dr. Floyd Thompson and several Associate Directors and Division Chiefs, Associate Director Ed Kilgore turned to Associate Director Clint Brown and said, "Damn it, let's do it Brownie." Clint Brown nodded. Dr. Thompson thought for a moment and also nodded in agreement. The rest agreed, and the decision was made.[2]

With that decision finalized, plans were put in place to have the Lunar Orbiter managed by Langley. A contractor would build the spacecraft. The contract called for five Lunar Orbiter vehicles to be built, with one Lunar Orbiter to be launched every three months. The reason for the five spacecraft came from earlier studies. In these studies, it was shown that, in order to have approximately an 80 percent chance of two successful flights, you needed to have five flights.[3] Those were the odds when we began the Lunar Orbiter Project.

Compared with that history, our 5 for 5 success rate of Lunar Orbiter is even more extraordinary.

12-Who Will Get The Job?

1963 Now Langley had to develop the Request for Proposals (RFP), the announcement that invites bids for an activity. Headquarters was pushing heavily for the spin stabilized spacecraft concept.[1] However, Langley held firm that the RFP needed to allow for a three axis controlled spacecraft concept as well as the spin stabilized spacecraft concept. Finally Langley, led by Dr. Floyd Thompson (Center Director), and the Lunar Orbiter Project Office headed by Cliff Nelson, prevailed. When the RFP went out for the Lunar Orbiter Project, the spacecraft concept was unrestricted, opening up the possibility for proposals for a spin stabilized or three axis controlled spacecraft concept. This was to be a very important decision.

The RFP for the Lunar Orbiter had gone out from Langley in August 1963, and the proposals came back in October 1963.[2] The Source Evaluation Board (SEB) went to work reviewing the proposals to determine which company would build the Lunar Orbiter spacecraft. After some investigation by the SEB, it was clear there were two front running bidders: The Boeing Company and The Hughes Corporation.[3]

Boeing had proposed using a satellite that was three axis stabilized and would maintain orientation in the cruise mode by sensing the sun and the star, Canopus. The spacecraft could then be maneuvered in roll, pitch, and yaw to any orientation desired so that the camera could be aimed at any point on the lunar surface. The spacecraft orientation could be aligned with the spacecraft motion over the lunar surface so the film could be locked onto a plate called a platen. The platen could be driven in only one direction while the camera shutter was open, compensating for the fact that the spacecraft was moving over the lunar surface. Therefore, the image would not be smeared.

Instead of having to maneuver the spacecraft to take pictures, Hughes had proposed using a spinning satellite. In this case, the satellite would snap the shutter on the camera at the time in the spacecraft's rotation when the camera pointed at the area of the Moon to be photographed. Because of the spinning motion and the fact that the spacecraft would not be lined up with its direction of travel, the apparent motion on the film would be much more complicated to calculate than in the Boeing spacecraft. This, if not corrected, would cause smearing of the pictures, just like moving a camera while taking a picture. The Hughes Corporation proposed to account for this motion by locking the film on the platen and moving the platen as necessary. They developed a simple mathematical formula to compute how the platen would have to be moved in order to avoid smear on the pictures.

Israel Taback, affectionately known as "Is," was the chief engineer for the Lunar Orbiter Project. "Is" was the best chief engineer I have ever worked with. He always had the uncanny knack of asking just the correct questions, and he did it this time. His question was, "Is it really that simple? Will this simple equation adequately describe all of the motions you will need to compensate for in order to avoid smear?"

The call came to Bull that some of the Source Evaluation Board members would like to meet with him and a few others to examine a question with regard to the Hughes proposal. To this day I do not know why, but Bull looked around and said, "Come on, Newk. We have a meeting to attend."

Off we went to the meeting where "Is" explained the question, and possible methods of attacking the issue were discussed. I remember one fellow saying, "One thing is clear, however you solve this thing, there will be a whole lot of algebraic bulldozing." I was to find that out very soon. As we walked back to the office, Bull looked over at me and said, "Okay, Newk, it's all yours."

That was the beginning of my algebraic bulldozing. The problem, simply stated, was to mathematically describe the movement of a point on the lunar surface as it would be seen by the film in the camera as the spacecraft spun and traveled over the lunar surface. In other words, think about the smear you would get on your picture if you were spinning around while running across the yard taking pictures at various angles to your spin and your direction of running.

Now mathematically model the movement so that you can move the film in the camera while the shutter is open to compensate for the motion. Also, realize the shutter stays open for a relatively long time, because we have to use slow film so it will not be damaged by solar radiation.

I started to develop the equations to describe the motion. I quickly moved to the two foot by three foot paper that goes on a flip chart. After outgrowing this, I went to three foot wide paper on a roll and used a conference table. Then there were two conference tables end to end for the general equations and two more conference tables for the mathematical description of the terms.

Finally, late one Sunday afternoon just before the questioning session with Hughes, I finished the derivation of the motion. I ran a few quick checks to assure there were no errors and the equations were solid. I did a few more analyses and proved that the Hughes equation would not work. In fact, the motion was so complicated that it would need two fast moving, computer controlled , piston like devices and a rather large computer to drive the platen. I wrote on the board, "I solved the damn thing", and went home for the evening.

Prior to the Source Evaluation Board (SEB) recommendation, both Hughes and Boeing were brought in for final questioning. This was a very formal process. The initial set of questions and answers were being recorded.

In the Hughes questioning, I was next to "Is" with a prepared set of questions. When it came time to ask about the platen motion problem, "Is" asked, "Is the equation in the proposal sufficient to avoid smear on the photographs?"

The only answer Hughes could give was, "Yes".

"Is" responded, "It is obvious that the equation in the proposal is an approximation. Have you ever developed the actual equations?"

The answer was "No, we have not. The mathematics are horrendous. We have attempted it several times and failed. We have not been able to complete the derivation."

Well, my hat size went up a couple of notches.

I heard "Is" say, "We have a fellow here who has developed the equations, and he will show you how much of an error you have in your approximation, and will also show you the completed development."

For the next two weeks, I had several of the Hughes Company division chiefs, vice presidents, and technical leads standing on my shoulders trying to find a mistake in my work. They also phoned the equations back to Hughes for the technical troops there to examine and to program. This was serious business. If they could not find an error in my derivations, Hughes was completely out of the competition, as their concept wouldn't work.

We performed many calculations of different conditions by hand in order to show the Hughes folks how bad the smear could be. I overheard one of the Hughes individuals on the phone back to headquarters saying, "I tell you I think they got it right! They got this young kid who derived the full set of equations, and they got this fat guy who can get six damn decimal places with this two foot long slide rule, and damn it, we cannot find an error!"

Jessie "The fat guy"
Later in Lunar Orbiter Operations.

The fat guy was Jessie Timmons. Jess was one of Buglia's Boys, and he could get an amazing amount of accuracy on that blasted slide rule (the engineer's computer before computers), partly by intuition and mental computations.

Sometimes he just guessed if he wanted to impress, and I am sure that is what he was doing in this case. In any case, he was much more accurate than needed, so typically no one questioned him.

In the final analysis, the Hughes folks became convinced I was right and began to admit defeat. This immediately put them out of the running for the Lunar Orbiter contract. In May 1964, Dr. James Webb, NASA Administrator, signed the contract with The Boeing Company. The Lunar Orbiter Project, the last of the three missions to characterize the lunar surface, was under way.[3]

13-The Rest of the Story

A few weeks had passed since the announcement that The Boeing Company was the winning bidder. Boeing would build the five Lunar Orbiter spacecraft and perform the flight operations.

We were all settling back into our routines. We were picking up a lot of work for the Lunar Orbiter Project Office (LOPO). I was working at my desk when a call came in. The caller identified himself and said he had a client who wished to offer me a job. The client would make himself known if I was interested.

Of course, I figured the client had to be Hughes. No one else except Langley knew I existed. "What type of salary are they talking about?" I asked. The caller quoted a number that was three times my existing salary, but he said it was negotiable. His client would consider going higher if I wished.

Well, if someone offers you three times your existing salary, and that is the starting point of the negotiations, you have to listen. I admit he got my attention. I told him I would like to think about it for a few days and get back with him.

I did think very hard about the offer. Hey, wouldn't you? However, in the end I turned the recruiter off and turned the offer down.

I did this for one main reason:

We Were Going To The Moon!

Hughes was not. They would be looking at the mission from the sidelines and reading the headlines with the rest of humanity. Because of my recent work in exposing the error in the Hughes design, I was going to play an important role in the Lunar Orbiter mission. I did not know what it was yet, but I knew it would be significant. I wasn't going to turn down that opportunity.

14-Some Issues Never Die

The question of the spinning satellite concept simply would not go away. Even after the winning proposal was announced, Congressman Earl Wilson of Indiana questioned the results because of the major difference in cost between The Boeing Company and the Hughes proposals. Boeing's proposal was approximately twice as expensive as Hughes'.[1]

Again, we recounted the argument that the image motion on the film could not be controlled with a spinning spacecraft if a low speed film was used. In addition, Dr. Trutz Foelsche of Langley had shown that a high speed film would very likely be damaged by radiation, because you simply could not provide enough shielding.[2]

Once again, it was shown that Hughes was caught between the two blades of the scissors, one being the incapability of performing the image motion compensation for slow film, and the other being radiation damage that would occur if high speed film were used. Therefore, the idea of a spinning spacecraft to photograph the lunar surface simply would not work.

15-The Lunar Orbiter Team

As the question of Langley taking on the project began to take shape, a small management team for the Lunar Orbiter Project was formed. Heading the team was Cliff Nelson.

Cliff was a mild mannered soft spoken fellow, but one who commanded respect. Israel Taback was the Chief Engineer. Whenever I went to lunch with "Is," I always made sure I had a pen and a few 3 by 5 inch cards in my pocket so I could take notes. Other team members were John Graham, who had been head of the recovery operations for Project Mercury; Eugene (Ed) Brummer, in charge of the spacecraft development; William (Bill) Boyer, who would head up operations; and Norman (Norm) Crabill, who would head up the mission design effort. Dr. Floyd Thompson brought in Jim Martin as Deputy Project Manager (later to become the Project Manager for the Viking Project) to support Cliff in the business management aspects of the project.

While our group (Buglia's Boys) would not be a physical part of the project office, it was becoming clear that we would be totally committed to the project. We would be responsible for the technical support needed to design the Lunar Orbiter missions and for placing major requirements on the spacecraft. These requirements included fuel requirements (given the possible dispersions in the translunar trajectories), lunar orbit insertion maneuvers, and lunar orbits. We would also pick up responsibility for overseeing the computer programs that would design the mission, compute the velocity maneuvers, and the photographic maneuvers occurring during flight operations. We had a good set of analysts: George Young, Richard Green, Jess Timmons, whom I mentioned earlier, and John Unangst. Yet we needed someone who really understood statistical analysis. I went to Bull and George and said I had the perfect person, Larry Hoffman. I had gone to high school with Larry, and he was as close to a mathematical genius as I had ever met. In high school, when the rest of us were reading dime novels, Larry was reading calculus books for fun. He scored in the top one-half of one percent of the nation on the college boards.

He did not apply to colleges. They applied to him. After graduating from Tulane University with a degree in (what else) mathematics, he was now working at Langley in another area. There was no contest. All parties agreed, and Larry became one of Buglia's Boys.

It was agreed the Jet Propulsion Laboratory (JPL) in Pasadena, California would furnish the Deep Space Net (DSN). The DSN consisted of the tracking stations to track the Lunar Orbiter, receive the down link transmissions, and transmit the commands. JPL would also furnish the Space Flight Operations Facility (SFOF) to control the mission.

JPL also agreed to conduct a space flight school that Langley and Boeing people could attend to learn how these large space missions were run. Joe Brenkle at JPL arranged for the school and gave us access to as many experts in the various fields as we requested. I was sent from Langley and spent about six weeks, absorbing as much knowledge as possible. It was a truly great experience for a young fellow just two years out of college.

Lewis Research Center (now the Glenn Research Center) would be responsible for the launch vehicle, the Atlas, with an Agena rocket second stage.

The Boeing Company would be responsible for the development of the five Lunar Orbiters (LO), the integration of the LO with the launch vehicle, and the mission design. Boeing would also be responsible for using the tracking data and determining the spacecraft trajectory as well as the development of all of the operations software. These programs would compute the velocity and photographic maneuvers, and all associated software that would monitor and command the spacecraft.

A well respected program manager named Robert Helberg headed the Boeing team. He was a distinguished looking, white haired gentleman known as the Silver Fox. He and Cliff Nelson, NASA Langley Project Manager, made a great team to bring two of the main players (Langley and Boeing) together. Later, there was to be a crater on the Moon named Crater Helberg, in Bob's honor.

The Office of Space Science and Applications (OSSA) was responsible for the program at Headquarters. Dr. Homer Newell was director of OSSA. Deputy Director was Edgar (Ed) Cortright, who would later become the Director of Langley. Lee Scherer was the Program Manager, with Leon Kosofsky as the Program Engineer

16-The Boeing Design

The Lunar Orbiter Spacecraft
Credit: NASA

The Boeing design for Lunar Orbiter (LO) was a spacecraft wrapped around a photographic subsystem. I have included a schematic diagram with numbered descriptions to help identify the various parts of the spacecraft on the following page.

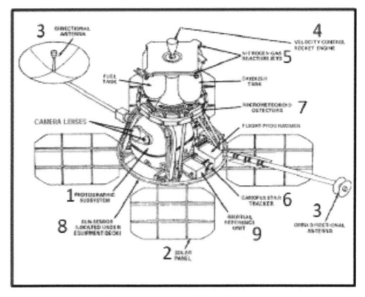

Schematic of the LO Spacecraft
Credit: PD USGOV NASA

1.   The photographic subsystem, shown with the two camera lenses, was contained inside a sealed unit that looked like a squashed sphere. The large camera lens was for the high resolution camera, and the smaller lens served the medium resolution camera.

2.   The four paddles were the solar panels.

3.   The omni directional antenna would be used for commands and for getting data on the health of the LO. The directional antenna would transmit the photographs back to Earth.

4.   The velocity control rocket engine was at the top of the spacecraft, and the fuel and oxidizer tanks were just below it. This was the engine that would deboost the Lunar Orbiter into lunar orbit and make any maneuvers needed for trajectory changes.

5. The nitrogen gas reaction jets are shown at the corners of the top plate and would keep the S/C pointed in the correct direction and maneuver it to whatever orientation was needed.

6. The Canopus Star Tracker was located so it could see and lock on to the star Canopus. With the sun sensor locked on to the sun and the Canopus Star Tracker locked on to Canopus, we would know the exact orientation of the spacecraft.

7. The micrometeoroid detectors were simply pressurized cans. If a micrometeoroid punctured the can, the pressure would drop, and the sensor would record the drop that would indicate a hit.

8. The sun sensor was located under the equipment deck, so that it and the solar panels would both be looking directly at the sun.

9. The inertial reference unit consisted of a set of gyroscopes. When we needed to go to a particular orientation (such as for the deboost maneuver to get into the orbit about the Moon or to orient the Lunar Orbiter properly for taking pictures), we could command the Lunar Orbiter to go to that orientation. The gyros would know when we got there, be able to provide information to the Lunar Orbiter computer so it could command the attitude control rockets to maintain orientation for as long as we needed, and then return back to our Sun Canopus reference.

17- The Camera

Lunar Orbiter with Camera Door Open
A close up picture of the LO camera is shown above.
In the picture, the camera thermal door is placed open,
showing the high and medium resolution lenses.
Credit: PD USGOV NASA

The camera system that we used was designed after a spy satellite camera that the Air Force had been using to take pictures of strategic locations from orbit. They, as well as we, were interested in obtaining the highest resolution possible, so it was an obvious choice. In the Air Force version, the camera and spacecraft system would take the pictures, and then the film canister and camera would deboost out of orbit, fly into the earth's atmosphere behind an aeroshell, and deploy a parachute. During its parachute descent, it would be snagged by an Air Force plane and the film taken back to a base and developed.

In our adaptation of the camera system, the film was released from the film supply reel and flowed across the high and medium resolution film platens, where it was secured as the pictures were taken. After the film had been exposed, it would move around a processer drum and come in contact with the developer, "Bimat," which was being fed from another set of reels. The film would pass by a heater to dry, and the developing process would be stopped.

The developed film, which was now a normal negative of a picture, proceeded to a scanner that scanned different lines across the film by shining a very thin beam of light through it. The different shades of gray on the film allowed more or less light to travel through the film, and these different light intensities were measured by the photocell on the other side. The resulting signal was amplified and sent down to Earth, and there the reverse process was performed. A thin beam of light was used again, and its intensity was modulated by the signal from the spacecraft.

This light beam scanned across the new film and exposed it to the varying shades of gray, just as those in the original picture. These strips of film were put together on a light box to form the picture taken by the LO camera. The resulting picture that was now on the light box was photographed to produce the final Lunar Orbiter picture. This is why the LO pictures always have the lines in them, since these lines are the edges of the filmstrips that were used to reconstruct the photographs.

John Graham
(Head of the Photographic Ground Data Systems)
Shown putting film strips on the light box.
Credit: NASA 1967 L07069

18-What Type Of Mission Was That?

When the original Request for Proposal (RFP) went out, it was based on what we termed a "concentrated mission." In this mission, a particular area would be chosen as a possible Apollo landing site, or area of interest, and the Lunar Orbiter would take pictures of that area on successive orbits until the film was used up. As this was occurring, the film would be advanced, developed with the Bimat, and moved on to the readout area.

As the Lunar Orbiter Project got under way, there were many meetings and discussions regarding how to use the Lunar Orbiter photography to maximize the benefit to Apollo. During this time, Norm Crabill had hired a new fellow named Tom Young. Norm would later say, "Tom was the brightest guy he had ever met."

Norm and Tom went to Flagstaff, Arizona to talk with the US Geological Survey (USGS) personnel, who would play a major role in the selection of Apollo landing sites. They met with Larry Rowan, who was to head the USGS effort to support the selection of the LO photographic sites. Larry and Hal Masursky of the USGS had been thinking about this for a long time. Larry met with Norm and Tom, surrounded by huge mosaics of the lunar surface. He began to point out different sites that could be possible Apollo landing areas, and these were spread all over the Apollo area of interest from +/-5 degrees latitude and +/-45 degrees longitude.

Then the light dawned. Norm and Tom looked at each other and quickly acknowledged that they had a completely new mission design on their hands. They could not afford to take all of the pictures and use all of the film available on each Lunar Orbiter mission on just one area. What if that area was not suitable? Wouldn't it be better to get a look at many areas and then have a better set of choices for the Apollo landing sites?

It became clear to both of them that the best mission design for LO was not the concentrated mission where all of the film was used in one area. Instead, the film would be used more judiciously with pictures taken of a higher number of promising Apollo sites. These sites would be spread out all over the Apollo area of interest. This became known as the "distributed mission."

Once the idea of the distributed mission was presented, it was obvious to all involved that it made sense. That was the mission the Lunar Orbiter should undertake. Only one problem remained. The spacecraft and camera system were designed for the concentrated mission. Since the Apollo landing sites were spread throughout the Apollo area of interest, it meant that the Lunar Orbiter would take some pictures and wait while the next Apollo site came under the orbit track. The LO would then take pictures of that site, resulting in a few orbits in which no pictures were taken. This meant that the developer (Bimat) would be in contact with the same film for longer periods of time. It would dry out, stick to the film, and cause the image to be spoiled. There was a concern that it could cause the film to tear or jam the entire system, rendering the camera useless. This would also require the individual LO missions to be approximately a month in length, longer than the several days originally planned.

We now had the correct and proper mission concept and a spacecraft that was not designed to fly that mission. The discussions began in earnest. Boeing's position was, "This was not the mission you asked us to bid on. You are asking us to put ourselves at risk of losing our incentive because you want to do something different. This does not make sense. You are asking us to increase our risk of not completing the missions."

That was the key word: risk. Was the distributed mission really riskier than the concentrated mission? Norm gave Tom Young the problem. "I want you to find out how risky this distributed mission is. What is the difference in risk between performing the concentrated mission and the distributed mission?"

Norm would say later, "I thought, 'That would keep him busy for a long time, but he showed up at my desk in three weeks with the results.' They were totally verifiable, and they showed the difference in the risk between the two types of missions was negligible." The major adjustment to the mission in order to keep the risk down would be the requirement that a picture be taken at least every eight hours in order to move the Bimat so it did not contact the same film for longer periods of time. This would keep it from sticking to the film.

Even with this proposed fix, Boeing was not happy. However, they did not have any real argument at that point. Armed with this data and after several more meetings, Langley was able to convince Boeing to commit to the distributed mission and begin to concentrate on the final preparations for launch.

19-The First Of The Lunar Orbiter Missions

YOU CAN'T FIND WHAT?

NASA had successfully completed the last three Ranger missions and taken the photographs as the spacecraft headed for an impact with the Moon. The Russians had conducted the first soft landing on the Moon with Luna 9 in February 1966. They had placed Luna 10 in orbit around the Moon in April of 1966. With Luna 10 Russia had gained information on the lunar gravitational field but had taken no photographs of the surface.

Now it was Lunar Orbiter's turn. We would launch from Cape Canaveral and control the missions from the Space Flight Operations Facility (SFOF) at the Jet Propulsion Laboratory in Pasadena, California. Some of the project members headed for the Cape, while the rest of us headed for California. We would launch one of the five LOs every three months, starting August 1966.

The missions would last for about a month. I would be at the SFOF supporting each mission. In between, I would be traveling to various places getting ready for the future mission and returning to the SFOF for the next mission. Oh, and I would be making a few occasional trips back home to Gloucester, Virginia to see my bride, Peggy.

This was to be our first launch of the five LOs. Excitement was high on both coasts. As the launch team readied for the launch at the Cape, we were preparing in the SFOF to begin the tracking and computation of the first midcourse maneuver to correct the residual errors that occur in any launch.

LO 1 Blasts Off for a Rendezvous with the Moon
August 10, 1966
Credit: NASA L 66 6250

On August 10, 1966, the Atlas Agena launch vehicle performed flawlessly, and LO 1 began its journey to the Moon with a planned arrival in approximately 92 hours.[1] After many months of developing the LO, readying the Space Flight Operations Facility (SFOF) at the Jet Propulsion Laboratory and the tracking stations that comprised the Deep Space Net, plus developing the flight operations software that would be used to navigate and control the mission, we were on our way. This would prove to be a rich learning experience.

The spacecraft deployed the solar panels and the antennas, and began looking for the sun. "We have sun lock," was the word from the Spacecraft Performance Analysis and Command (SPAC) area. Now we commanded the roll maneuver to acquire Canopus. "Canopus sensor voltage much higher than expected," was the next report from SPAC, meaning that the sensor was receiving more light than it should. After much analysis, they found the Canopus sensor was getting reflected light off various parts of the spacecraft.

This was a serious problem. We had to have the Canopus sensor looking at a recognized object so we would have a known orientation of the spacecraft, or the mission would be lost. We gave the LO the command to perform additional roll maneuvers to find what we might be able to see with the sensor.

"We have the lighted limb of the Moon," was the response from SPAC. We could use that in order to provide a known spacecraft orientation.

The LO was now locked on the sun and the lighted crescent of the Moon. That was not exactly as we had planned, but it would work.

It is well established that you do not adjust the command generating software once it has been tested and certified. But in emergencies you have to improvise, and this qualified. (The light reflection problem was at least partially corrected for the rest of the LOs by painting the reflective surfaces with a special coating.)

These anomalies, plus others to follow, did not damage the mission – but they sure kept the operations team busy.

20-Bouncing Around

LESSON NUMBER 1

MISSION DAY 1 AND 2

Now began the calculations and analyses to determine the proper midcourse maneuver in order to correct the small errors that always exist in a launch sequence. The JPL tracking data analysis team was doing a great job of using the three deep space tracking stations in Woomera, Australia; Madrid, Spain; and Goldstone, California; and providing a good clean set of tracking data.

Using this data, Matt Grogan and the Boeing Navigation Team began determining the actual trajectory. Dale Shellhorn and his team began to compute the first midcourse maneuver in order to arrive at the proper position at the Moon for lunar orbit insertion.

The NASA team of navigation mission advisors, Bob Tolson and John Gapcynski, led by Bill Michael, was also reducing the data and following the Boeing orbit determination process. The mission design team of Woody Lovelace, Larry Hoffman, John Unangst, and me made our own calculations and followed the maneuver design. It was decided that the midcourse maneuver would occur at 40 hours into the flight, and would be approximately 19 meters per second.

One of the initial items in the sequence for the midcourse maneuver was to open the lines so the fuel and oxidizer could reach the velocity control engine valves. These lines were normally closed for safety until the propellant was needed. They were opened by firing a pyrotechnic device (called a "squib") that ruptured a diaphragm and let the fuel through. After the squibs were fired, the propellant could go through the lines and reach the velocity engine control valves. These valves could then be opened and closed by commands given by the onboard computer.

We had a stored sequence of events that the computer used to conduct the maneuver. One of the first items in this sequence was to open the propellant lines. Then began the maneuvers to orient the LO for the midcourse and fire the engine. Immediately when we fired the squibs, SPAC reported, "Lost orientation, lost sun and lunar limb."

The next order: "Abort midcourse immediately." We were young at this business. We had not considered that the fuel, which was pressurized, would flow quickly through the lines to the closed valves. This would cause a transfer of mass in the LO. Since the LO was not held in place, it would change orientation...and it did.

After all the analysis was over and the conclusions were reached, we reacquired sun and the lunar limb and computed a new midcourse for 7 hours later. This was performed perfectly, and we were off to the Moon again.

21-The Data Analyst

The first midcourse was over, and we were waiting for the tracking data to be analyzed to see if everything was continuing to go as planned. It was late into the night, a quiet time in the SFOF. For the first time in the mission, I had a chance to sit back and realize where I was and what I was doing. I was in an operations center, computing maneuvers for a spacecraft, and participating in a mission to the Moon!

I looked around a moment just to soak in the sight and the let the realization set in. I looked at the Spacecraft Performance Analysis and Command (SPAC) area and saw the SPAC guys looking over the stream of data coming in to make sure the LO was continuing to operate properly. I saw the Flight Path Analysis guys beginning to look at the earlier tracking data and consider the next maneuver we would have to make.

I looked at the Data Analysis area and watched as the different data analysts were reviewing the incoming data on the Teletype machines. One was watching the Teletype very intensely and talking on the phone. Since Woomera was the receiving station at that time, I assumed he was verifying that the data was being transmitted properly.

While I was basking in this moment, realizing that I was involved in a pioneering mission, I saw the data analyst put his lips to the phone and give it a long, juicy kiss.

Obviously he wasn't talking to Woomera!

22-Roll, Pitch, Bang

LESSON NUMBER 2

After another midcourse maneuver came the deboost maneuver to place the LO spacecraft into orbit about the Moon. All maneuvers were calculated, the commands sent, and the onboard computer had timed down to the maneuver time, performed the spacecraft attitude maneuvers, and fired the deboost engine..

All had gone well. It was then we realized that we did not do things the way the experienced space mission guys did it.

Dr. William Pickering, who headed the Jet Propulsion Lab (JPL), had brought in a delegation to witness the maneuver and the determination of the subsequent orbit. Pickering was shown the command sequence.

"Is this how you really do this maneuver?" Pickering asked.

Bill Boyer, who was head of mission operations, said, "Sure. Do you do these maneuvers differently?"

"We certainly do. We roll the spacecraft, wait ten seconds for the attitude control oscillations to damp out, pitch, wait ten seconds, and then fire the engines," Pickering responded.

We had done what Bill called a roll, pitch, bang, and we had not given the control system oscillations a chance to damp out. Said another way, we had not given the Lunar Orbiter time to stabilize after the roll maneuver or the pitch maneuver before firing the engine. We were lucky. The roll, pitch, bang maneuver had worked fine this time, but we were learning how these big space missions were run.[1]

23-Mission Design

MISSION DAY 5 AND BEYOND

After performing the deboost maneuver we went into an orbit inclined 12 degrees to the Lunar equator with an apolune (point of highest altitude above the lunar surface) of about 1100 miles and a perilune (point of closest approach to the lunar surface) of about 120 miles. We would stay in this orbit for a few days in order to determine the precise orbital parameters and attempt to learn more about the lunar gravitational field. Then we would deboost again to bring the perilune down to approximately 30 to 35 miles above the lunar surface.

With the descent to the lower orbit we would begin the main photographic mission to photograph the Apollo landing sites.

From this lower altitude we would take a single set of pictures or sequences of eight or sixteen sets. Each time the camera shuttered it would take a high and a medium resolution picture set. This resulted in the typical photographic footprint for a single high and medium resolution picture as shown on the next page.

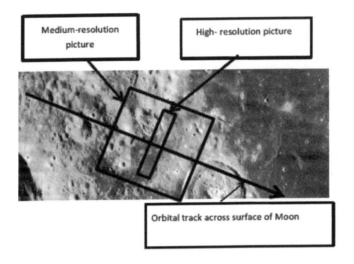

Lunar Orbiter Picture Format on Moon

Two actual photographs from a set of medium and high resolution frames are shown on the next page to give the reader a feeling for the difference in the detail between the high and medium resolution pictures.

- The first photograph is part of a medium resolution frame.

- The second photograph is the high resolution frame taken at the same time and located inside the box.

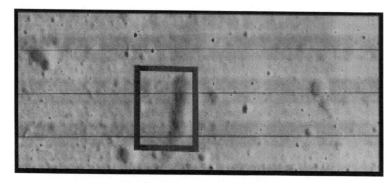

Medium Resolution Frame
Box showing position of high resolution frame.
Credit: NASA November 19, 1966, Orbiter 2

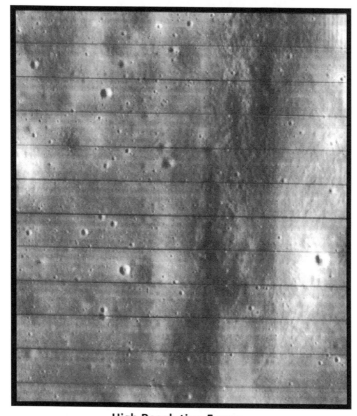

High Resolution Frame
Credit: NASA November 19, 1966, Orbiter 2-5

It is interesting to see areas that look reasonably smooth at medium resolution turn out to be quite a bit rougher at the higher resolution. We were to discover this later as well, when we looked for landing sites for the Viking Landers on Mars.

The mission design for LO 1 (which is typical of the first three missions) is shown below. We would now be in an orbit with perilune near the equator and with the initial photographic orbit stationed to the right hand side of the Apollo area of interest, (region for possible selection of Apollo landing sites).

The small black rectangles within the area of interest represent the ten different sites we were to photograph during this mission.

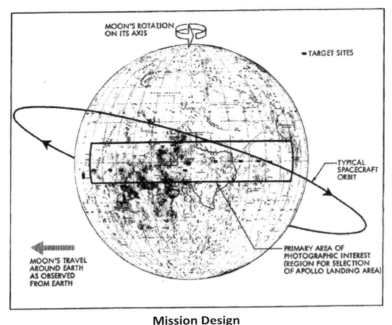

Mission Design
Moon with Spacecraft Orbit and Apollo Target Sites
Credit: Reference 16: Page 4

In the beginning of each of the LO missions, the front face of the Moon would be partially illuminated and look like a new Moon from Earth. We would then take pictures as the Moon rotated under us. Since the orbit was stationary in space and the Moon was rotating under the orbit we could take pictures of the possible landing sites as they came into view. At the end of the mission the orbit would be over the left side of the area of interest and the photographic mission would be finished with only the remaining pictures to be read out.

As Tom Young used to say, "Take heart everyone; we go home on the full Moon."

24-The Newcomb Maneuver

MISSION DAYS 9, 10, 11

After we brought the LO 1 down to the main photographic orbit with a perilune altitude of 35 miles and began taking pictures, we realized we had a problem. The medium resolution camera was providing great pictures. However, the high resolution camera had smear in the pictures. We learned that a sensor that was supposed to correct for the smear was not working.

There was some indication that if the LO was brought down to a lower altitude, the faster speed over the surface might trigger the sensor to work better and correct the smear. We decided to perform another maneuver in order to bring the LO down from a perilune of 34 to 24 miles.

Cliff (The LO Project Manager) placed me in charge of working with The Boeing Company in the design of this maneuver. It came to be known as the Newcomb Maneuver. It was a long 36 hours. We knew we were going to get pretty close to the Moon, and it would not take much of an over velocity to cause the LO to crash into the lunar surface. Therefore, this maneuver had to be calculated very precisely, considering all error sources, so that the sum of those possible errors would not cause a problem.

Working with Dick Rudd and other members of Dale Shellhorn's team from The Boeing Company, we calculated and recalculated the maneuver many times. Also, the navigation team, headed by Matt Grogan for Boeing and Bill Michaels for Langley, was continually coming up with updates on the spacecraft orbital parameters. The Spacecraft Performance Analysis and Command Group were reviewing the possible spacecraft errors that might be introduced. It was a total team effort to make sure we got it right.

On August 21, LO made the velocity maneuver. We waited for the navigation guys to tell us the results. It seemed like days. In reality, it was only a matter of about 20 minutes when they completed analyzing the first tracking data – and we nailed it. We were shooting for 24 miles altitude, and we got 24 miles altitude.

After the maneuver accuracy was confirmed, I went home to the apartment where Peggy and I were staying and crashed. The sleep was wonderful after the 36 hours of tension and calculations.

My cherished sleep did not last long. Peggy woke me up after three hours and said that Cliff had called, and I was to be in a press conference.

He said, "Get to the Space Flight Operations Center (SFOF) as quickly as possible."

After stumbling around trying to get rid of a 48 hour beard, finding clothes, and making notes as Peggy drove me, we arrived at the SFOF.

"Are you ready for the press conference?" Cliff asked. "I want you to explain the maneuver we just made."

"I guess. How soon is the conference?"

"In five minutes. We need to leave now."

So, with my notes and some fast consultation with my Boeing friends, I headed off to the press conference to explain the maneuver.

The Press Conference
I am to the far left of the picture, trying to figure out what I am going to say. From my left to right, Lee Scherer, Program Manager; Dr. Floyd Thompson, Langley Center Director; Cliff Nelson, Project Manager; and Israel Taback, Chief Engineer
Credit: NASA 1967 L 02769

Technology Week, August 29, 1966, had this to say about the press conference and the maneuver:

> *"A highly successful deboost maneuver into low orbit took place at 2:50 a.m. (PCT) on August 21. We are now thoroughly satisfied with the position of the vehicle for the mission for which it was designed and prepared,"* commented Dr. Floyd L Thompson, Director, NASA Langley Research Center."

I described the orbital parameters for which we were shooting, what we acquired, and the sensitivity to possible errors.

Again, *Technology Week*:

> *"The transfer maneuver was the most sensitive to errors of any maneuver we have performed yet," said Newcomb.*

Technology Week also reported,

> *"The tracking experts decided that the spacecraft perilune had been put within a city block of where it should be."*

Describing theManeuver
Credit: NASA 1967 L 02773

It turned out that the lower perilune did not solve the high resolution camera smear problem.

Fortunately, the medium resolution pictures were of very high quality and answered many questions regarding the Apollo landing sites.

The rest would be answered on LO 2 and 3.

25-The Forgotten "Picture Of The Century"

During this part of the mission, we realized we had another problem that would ultimately cause us to take some of the most interesting pictures of the LO missions.

You'll recall that, after the pictures were taken, they were processed by carrying them over a processor drum in contact with the developer called Bimat. We had originally believed that we would have to take pictures every four orbits in order to keep the Bimat from drying out and sticking to the film. We were now getting pictures back with spots indicating that the Bimat was drying faster than we had planned.

We needed to take additional pictures every orbit. There was much discussion between Cliff Nelson and Bob Helberg. Bob wanted to keep the camera thermal door shut and advance the film. NASA and Cliff did not want to give up those pictures and just waste the film.

"As long as we are going to advance the film, we should pick a target and take a picture," was Cliff's position.

"We are simply increasing the risk of not meeting the Apollo requirements, and therefore we are increasing the risk of Boeing not getting its proper incentive," was Bob Helberg's position.

These discussions came to a head when Bob Helberg realized

that there were suggestions floating around that we take a picture ofEarth.

Now Bob really dug in hisheels.

"Now you not only want to take extra chances in maneuvering the spacecraft, but you want to point the camera away from the Moon and take a picture! Then after we take a picture, we will have to reorient the spacecraft, and then we go into occultation behind the Moon, lose the signal, and hope we can reacquire it when we emerge! This is just too much risk for us to take! And if I hear about anybody planning this, I will personally fire them!"What Bob Helberg did not know was that Dale Shellhorn, head of the Flight Path Analysis and Command Group (FPAC) had already begun thinking about an Earth picture and was the one who had suggested it in the first place. He had mentioned it to Jim Martin, the Deputy Project Manager, and initially received a resounding NO! Finally, after a much more enthusiastic discussion about the idea, he had gotten Jim's approval to quietly investigate the possibility. He had been performing some calculations to see if it was feasible. All of this would come out later.

Then came the meetings between Floyd Thompson, Director, Langley Research Center; Lee Scherer, Program Manager for LO; Cliff Nelson, Project Manager; and Bob Helberg, Program Manager for Boeing. Finally Cliff Nelson took Bob Helberg aside and discussed thesituation.

"Bob, if I were on the award fee panel for the LO contract, and I realized that Boeing took the Earth picture, the first picture of Earth ever taken from deep space, and it turned out beautifully, that would be worth a lot of award fee money that you otherwise might lose because of some of the small problems we have been having. It would be worth a lot of fee, even if it caused some problems in obtaining some of the Apollo pictures."[1] (Of course Bob must have known Cliff was on the award fee panel.)

Finally Cliff said, "The picture is worth the risk."[2] Bob yielded, the agreements were reached, the decision was made, and the idea of taking the Earth picture was now in the open. Now we could begin to determine openly how we would implement it.

We had another problem. The software had not been designed to compute the maneuvers necessary to point the camera at targets other than those on the Moon. So we were going to have to "jury rig" the photographic maneuver program to compute the needed pointing angles.

Cliff came to me and said, "I want you to be the NASA guy responsible for working with Boeing and designing the Earth picture."

Dale Shellhorn, leader of the Boeing Flight Path Analysis and Command Group (FPAC), took the lead from the Boeing side and, along with Bill Moyer and John Hoos from Boeing, we began working the problem. We had to fool the software into thinking it was looking at the Moon when, in reality, it was looking at the Earth.

Dale went back into the computer room and got the complete set of punched cards that represented the photographic maneuver program. The deck of cards was about two feet long. The cards used in those days were approximately the size of a business envelope. The set of holes punched in each card could represent an equation, instruction or data. All of these cards together would then represent a set of computations that would solve some mathematical problem, such as computing a deboost maneuver or determining the pointing angles for, and timing to take, a picture. These cards would be inserted into large, car size computers that would translate the instructions on the card into computer language. The computer would execute the program by carrying out all of the computations.

I headed off to get the American Ephemeris and Nautical Almanac that we could use to get the exact location of the Moon from the Earth. We could then reverse the direction and get the exact location of the Earth from the Moon. After a lot of analysis, we developed the equations and parameters we needed to change in order to fool the spacecraft. We found the locations in the program card deck where we would need to insert the new cards.

I punched out the new cards, and we replaced some of the existing cards with the newly developed ones. We ran the software offline and developed the maneuver sequence that we fed to the Spacecraft Performance and Command Group (SPAC) for encoding and sending to the Deep Space Net for transmission to the spacecraft. The actual picture was taken on orbit sixteen, August 23, 1966.

Then we waited. It seemed like years. Finally, we had confirmation that the picture had been taken and transmitted back to the DSN station in Madrid. Again we waited.

With all of the folderol about taking the Earth picture, we had failed to let the Processing Center know that we had taken it. The picture was transmitted from Madrid back to the Center in the middle of the night.

John Graham had led the development of the Video Processing System at the tracking stations and the Video Processing Center at the SFOF. It was his job, and the job of his people, to reconstruct the pictures as they were sent back from the LO through the tracking stations and then finally to the SFOF.

John was the only one working that night as the Earth picture came in. As he put the filmstrips together, he first saw the black of space. He thought, "This is strange, the guys must have screwed up a set of commands." Then he saw some of the Moon come into the picture.

"Well, at least they got something," he thought.

Then some white fuzzy stuff appeared in the black of space above the Moon. He thought, "This is really strange; we must have some problem with the Bimat." He considered reporting it immediately, but he figured he would keep putting the strips together and then report what he thought was a problem with the film.

As he put more filmstrips together – "Holy shit!" He realized that he was the first person to see the historic first picture of Earth taken from deep space! The white fuzzy stuff was part of the cloud cover surrounding our small planet.[3]

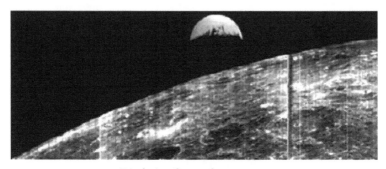

Earthrise from the moon
Credit: NASA 1966 L 07825

Now it was the mission advisors' turn to wait. It was later that one of John's men came out of the processing center and began quietly to put up a few strips of film that represented the left hand end of the picture on our light box. We just saw the black of deep space. Then some more strips came in, and they had some of the black of space and a sliver of the lunar surface. More filmstrips came. More black, more Moon.

More strips. And then the edge of a half lit Earth began to show! We had nailed it!

It was called "the picture of the century" and "the greatest shot taken since the invention of photography."[4]

Watching as the Earthrise Picture Unfolds
Shown left to right : Alton Mayo, NASA; Bill Michael, Head of the
NASA Navigation Team; Henry Holt, USGS;
John Newcomb; John Unangst, NASA; Terry Olefield, USGS;
Doug Lloyd, Bellcom; Tom Young, NASA.
From the body language of Henry Holt and me,
it appears that we have just seen the edge of the Earth.
Credit: NASA 1967 L 02601

Today Dale displays with pride the note from Jim Martin, the Deputy Project Manager for LO, congratulating him:

Jim's note to Dale
Courtesy of Dale Shelborn

Dale also displays the famous Earth picture on his wall in Tucson, Arizona. It is on a canvas in 32 inches by 87 inches format. He also donated a copy of the picture to hang on the wall of the Physics Department of his old alma mater, the University of Arizona, Tucson. I have a slightly smaller version, 22 inches by 48 inches on the wall by my office.

John Graham's copy of the picture hangs in the entrance to the Science Building at his Alma Mater, Randolph Macon University. The inscription reads:

> *"First Picture Of Earth Taken From Deep Space. Donated By Alumni John B Graham Who Constructed The Picture"*

The Earthrise picture also has another home in the Smithsonian.

Oh, and one further note. Bob Helberg would use this picture on his business card for many years.

Bob Helberg's business card
Courtesy of Dale Shelborn

In addition to being an amazing picture of earth, it showed how powerful the oblique photography was. We had concentrated, up to this point, on taking near vertical photographs in order to perform stereoscopic analysis to define slopes, bolders, and other features on the lunar surface useful to Apollo. Now we realized how much detail and relief we could see with the oblique shots. Many more were to come with future missions.

Fame is many times short lived. The team of Frank Borman, Jim Lovell, and Bill Anders, in December 1968 in Apollo 8, would take a beautiful color picture of Earth with the Moon in the foreground, a very similar design to ours. This became known as the "first" picture of Earth taken from deep space, and the black and white LO photograph, taken 2 1/2 years earlier, quickly sank into the dark corner of history.

Discussing the Earthrise Picture
Matt Grogan, who headed the navigation team (on left)
and Dale Shellhorn
Credit: NASA 1967 L 02576

26-Everything Old Is New Again

 As the Apollo Program continued and hundreds of beautiful color pictures and extremely high resolution black and white images came back from the Moon, the Lunar Orbiter pictures faded further into the background of everyone's memory.

However, one person would play a major part in bringing these pictures back to the foreground. The Los Angeles Times covered this extensively. Nancy Evans, a very determined woman was working at NASA Headquarters when she was given the job of "straightening out the NASA archives" involved with the missions around the Apollo time period.

Several years after the Lunar Orbiter missions had been completed, a clerk came to Nancy's office and wanted to know what to do with the storage room full of LO tapes that contained the photographs. (In many cases, after several years the old data is destroyed.) Nancy did not want these pictures destroyed and actually found a place to store them. Nancy realized that, in order to read the film, she would need the Ampex FR 900 Tape Drives, which were originally developed for the military and no longer in use. Over the years, she collected four of the drives and, having no place to store them, kept the one ton machines in her garage.

After approximately 20 years, she finally assembled a small team of space enthusiasts, including Dennis Wingo and Keith Cowing, and gained the support of Greg Smith, who was then the Deputy Director of the NASA Lunar Science Institute.

Pete Worden, the Director of the Ames Research Center, offered an old abandoned McDonald's as laboratory space. During the Apollo days, it had been called McMoon's, so that name seemed totally appropriate for the Lunar Orbiter Picture Reconstruction Project building.

With the help of an Army Veteran, Ken Zin, who had a lot of experience repairing top secret cryptographic machines, they were able to rejuvenate the old tape drives. These drives actually reproduced the LO pictures, not on strips of film as done originally, but as digitized images with much more resolution.

The pictures had a much enhanced dynamic range and no lines where the filmstrips had been glued together.

The first picture out of the starting gate, had to be, the earthrise picture from LO 1

As of this writing the team is still working, as funds are available, to restore the lunar pictures taken approximately 50 years ago. It turns out that these pictures are not just pretty pictures with higher resolution, but will be used to compare with more recent lunar pictures to determine if any changes on the Moon's surface have occurred in the intervening 50 years.

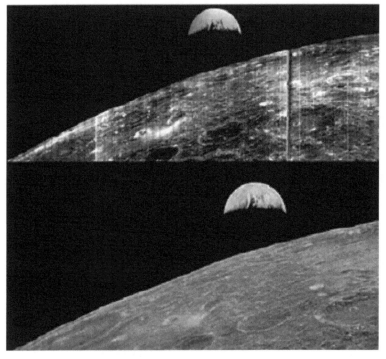

The Earthrise Picture Restored
The original (note lines) and restored picture
of the Earthrise over the Lunar landscape.
Credit: NASA/Lunar Orbiter Image Recovery Project

27-Kentucky Windage

The Kentucky long rifle was an extremely accurate firearm for its day, during the colonial period. It was muzzle loading and fired a round ball, but it had a distinct advantage over other, similar muzzleloaders. The barrel was rifled with grooves that caused the ball to spin. That gave it a much more stable trajectory. This rifle, placed in the hands of a trained marksman, was a lethal weapon. But the marksman himself had a particular technique that gave him an additional advantage.

If the marksman was firing at a target with a wind blowing from the side he would start by aiming the rifle precisely at the target and then he would move the aim point slightly to the side toward the direction from which the wind was coming. Therefore when the wind curved the bullet's trajectory slightly it would still hit the target.

This was called Kentucky Windage. And we had to use a version of this on Lunar Orbiter.

As we continued to review the photographs, we began to realize that all of the photographic sequences were resulting in photographs being taken down track from where we had intended. We checked all of our calculations, the trajectory prediction software, and anything else we could think of. Finally, someone suggested we compute the photographic maneuvers as normal and then just subtract some time from the computed shutter time. So we took the results from the sophisticated computational software, then applied our

Kentucky Windage by manually subtracting the appropriate amount of time and feeding that into the LO, and it worked. This solved the problem, but we still did not know what was causing the down track error.

Bill Sjogren, who was a Senior Research Scientist at JPL, was watching our attempts to get the proper timing for the photographs and trying to understand what was causing the problems. After a while, he realized there were anomalies in the tracking data when the LO flew over lunar craters that had been formed years ago by large meteors striking the lunar surface.

It was finally theorized that the impact of these meteors had melted some of the lunar interior. This heavy material had come toward the surface and solidified. This resulted in mass concentrations, now called mascons and these mascons had distorted the gravitational field and caused the lunar orbiter to speed up just a little as it came close to the lunar surface. This understanding gave us the cause of the problem, but we still did not know how to model the gravitational field properly to solve it. This same problem would also plague Apollo.

28-I Would Like To Introduce My Head Of Mission Operations

Before I start the following story, I must say that Bill Boyer and Cliff Nelson had worked together for over twenty years in one capacity or another. They had a lot of mutual admiration and respect for each other. This is one reason Cliff had chosen Bill to head up LO mission operations.

Now on with the story.

The congressional delegation had arrived earlier that day, and they were now making their way through the SFOF, led by Cliff Nelson. Cliff brought them through the Mission Advisors Area and introduced us to the group. We all shook hands, exchanged pleasantries, and tried to look very official. Then Cliff ushered them off and headed for the glass enclosed room in which Bill Boyer was carrying out his duties as Head of Mission Operations. Because this room was glass enclosed and next to the mission advisors' area, I could watch as Cliff introduced the delegation members to Bill. All seemed to go smoothly. Bill shook hands with the delegation members, more pleasantries were passed among the parties, and Cliff led the delegation on to other locations. Then I found out what really happened! Bill came into our area laughing like mad.

"I would like you to know, John Newcomb, that you are now talking to Bernard Jones."

"You are who? I am talking to who?"

"Bernard Jones. Cliff came in the Mission Operations Office with the delegation and started to introduce me. As he started the introduction, I saw his face go totally white. He could not remember my name! And he could not say, I would like you to meet my Head of Operations, oh, what is your name? So he did the only thing he could. He said 'I would like you to meet my head of operations, Bernard Jones.' So if you see me with the delegation, please remember to call me Bernard Jones."

"You better make sure the other troops know that."

"I am going to tell them that now." And off he went to put out the word. We had to make sure we remembered Bill's alias for the next few hours until the delegation left.

29-The Daily News

In every mission, there seems to be some unsung hero, or a set of unsung heroes, who will put together a periodic news bulletin giving the general non official information for the day. We had the LO BUGL. The BUGL would talk about the mission in a very informal way, but it would also spread any friendly gossip that might be interesting. Among all of the stress and work of the mission, we looked forward to this bit of relaxing chatter.

Due to a little bit of nostalgia, I suppose, I have saved a front sheet from some of the LO BUGLs all of these years. The one shown is the BUGL that came out when we were planning to photograph Surveyor that had landed on Oceanus Procellarum June 2, 1966.

The lines with the word "Zoom" in them indicate the LO flying over the Surveyor and taking the picture. However, with the high resolution photography being slightly smeared, we were not able to get a good picture of the Surveyor on this mission as we had originally planned. That had to wait until LO 3. However, this keepsake is a good reminder of the bulletin and the welcome daily chatter it contained.

For a while my wife, Peggy, made the news in the BUGL on a rather frequent basis. As I was leaving for the airplane for LO 5 , Peggy informed me that she and a friend of ours would drive out for the last mission. This meant that she would be driving completely across country during much of the LO 5 mission, which was a little cause for concern. I asked her to please wait until we completed the first midcourse maneuver so we would know the thing was working and then start.

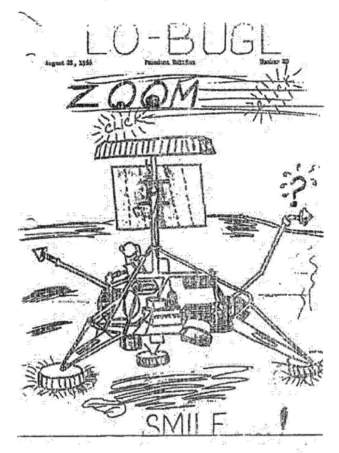

Front page of the BUGL, August 23, 1966

She had the car packed and ready and, after the successful maneuver, informed me that she was on her way. By this time, the entire flight team knew about the trip.

From then on, there was an evening phone call to establish her location and make sure she was okay. Of course, this "progress report" became part of the news in the BUGL, and the flight operations team kept up with her progress all the way across the country.

30-Moon And The Astronomers

It was around three a.m. It was quiet in the SFOF. Leon Kosofsky and I were watching the photos come back and get spliced onto the light box. Leon was the NASA Headquarters Program Engineer for LO. Leon was a geologist by education, but an astronomer at heart.

It was always a thrill to see the pictures emerge as the filmstrips were added. We had loaded several sets of commands, so there was nothing to do at this moment except enjoy the results of our labors. For a few moments, it was very peaceful to just watch the processing crew bring in the filmstrips, watch them being assembled, and the picture emerging.

All of a sudden, Leon broke the silence by saying, "Damn it, Newcomb. Do you know what you have done?"

I answered with a questioning "Huh?" I thought I had knocked over his coffee or something equally stupid.

"You have ruined the Moon for the astronomer."

"I have what? What are you talking about?"

"Before now, when the astronomer wanted to learn more about the Moon, he would go to the observatory, climb into the seat behind the telescope, freeze his tail off, and look at the Moon. It was cold. It was uncomfortable. It was exhausting. But it was romantic! Now when the astronomer wants to learn more about the Moon, where is he going to go? He is going to go to the library and look at your blasted pictures! Where is the romance in that?"

I have thought about this many times since that night. This was the first time the US had ever put a spacecraft in orbit around another celestial body and the first time the US had performed reconnaissance about any celestial body other than Earth. We were beginning to learn how to perform this reconnaissance, and we were proving the value of an orbiting spacecraft investigating another body.

Lunar Orbiter, in this way, ushered in the age of interplanetary spacecraft investigating other planets.

31-The Debrief

I had just come home from the LO 1 Mission, and this was my first day back in the office. Gus Sandahl, my Branch Head (my boss two layers up), came in and said that Hack Wilson, the Division Chief (my boss 3 layers up), wanted to see me and get a debriefing on the mission.

Hack was a great guy. He was a little gruff sometimes, but we all liked him, and we did not let the rough edges bother us. I headed down to his office to give my version of a debrief.

I was a little taken back by his approach.

"Well, Newk, did you have fun out in California and the big city of Los Angeles?"

"I did not get to see much of the city, since I was in the SFOF all of the time I wasn't sleeping."

"How was the beach?"

And then I lost it. "Listen, this mission was goddamn tough! If you think all I did was go to the beach and go touring, you're crazy as hell!"

With that, I stood up, banged on his desk, turned around, and walked out. As I walked down the hall, someone who had heard the commotion asked, "What just happened?"

"I just cussed out Hack Wilson."

"You what? Such a horrible end to a budding career."

"I don't give a damn!"

After I got back to my desk and regained my sanity, I finally realized what had just happened. I was still strung up tighter than a violin string from the tension of the last month. I never talked to people the way I had talked to Hack and certainly not to my upper management bosses.

In retrospect, I realized that Hack had heard about some of the things I had done, and he just wanted to talk to me and welcome me back from a very successful mission. He was trying to put me at ease, but it just so happened that his attempts at humor were the triggers that caused the tension in the gunpowder to explode.

I never brought up the incident again to Hack. I did not apologize. He never mentioned it to me. We got along fine from then on, and both of us seemed to put it aside like it never happened.

"Thanks, Hack."

We photographed thirteen sites for Apollo. One of the first sites to be photographed for that mission was Mare Tranquillitatis (the Sea of Tranquility).

It contained the impact point of Ranger 8. That was to be the area of the first manned landing on the Moon, where Neal Armstrong would step on the lunar soil for the first time and utter the famous words, "One small step for man; one giant leap for mankind."

During this mission, a discussion was brewing. It involved disobeying one of the Apollo flight rules again. One of the Apollo flight rules was that the extra frames we needed to take in order to move the film so the Bimat would not stick had to be pointed inside the Apollo area of interest. The reason for the rule was to use all film in support of Apollo and to avoid any maneuvers or actions that would endanger obtaining the pictures that Apollo needed.

However, Doug Lloyd (of Bellcom) had been arguing for a long time that we should take an oblique picture of the crater Copernicus. He had recommended this on several occasions, but NASA kept evoking the flight rule.

Finally, during the LO 2 mission operations, it was agreed that we could photograph Copernicus. This resulted in probably the second most famous photograph from LO after the Earth picture.

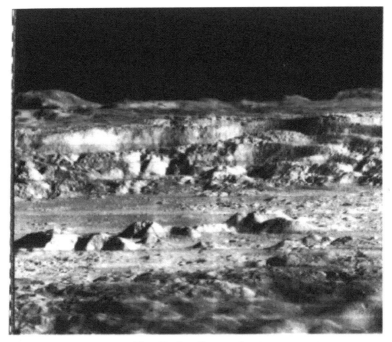

The Crater Copernicus
Taken from Lunar Orbiter 2
Credit: NASA 1966 L 09788

The crater Copernicus is 60 miles wide, and the mountains in the center are almost 1000 feet high. The picture gives a great feel for the enormity and grandeur of the crater as well as providing a wealth of useful information about the lunar surface. The news media described the Copernicus picture as one of the great pictures of the century[1]. We were learning with this picture and the earlier one of the Earth with the Moon in the foreground just how useful oblique pictures could be in understanding a foreign body.

We would use this knowledge in many of the remaining pictures we were to take on subsequent missions.

33-LO 3

LO 3 was launched on February 5, 1967 and, at the completion of that mission, we had met all of the Apollo photographic requirements. We had photographed all of the proposed Apollo landing sites.

Without a doubt, the most interesting picture from LO 3 was the photograph we finally got to take on February 22, 1967 of Surveyor 1 as it sat on the lunar surface in Oceanus Procellarum (Ocean of Storms) at 2.3°S and 43.°W.

Surveyor 1, launchecd May 30, 1966, landed on the lunar surface three days later on June 2. The image shows the Surveyor as a white dot (inside the circle) casting a shadow on the lunar surface.

As you can tell from the picture, it is not easy to see the Surveyor. In fact, we had photographed the area where Surveyor landed, and several of us were busily looking at the pictures, but we could not find the actual spacecraft.

Finally, the sharp eyes of Uriel (Woody) Lovelace (Head of the FPAC Mission Advisors) located the Surveyor, first by seeing the shadow and then deciphering the bright spot in front of the Surveyor as the reflection off the spacecraft.

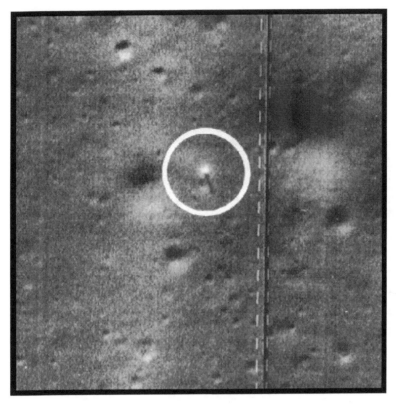

Surveyor
Kind of hard to see, but it is there.
Credit: NASA 1967 L 02934

Discussing the Next Maneuver
Bill Moyer, from Boeing; on my right, and George Lawrence, from
NASA; and John Hoos, from Boeing, across the table.
Credit: NASA 1967 ·L 05057

With the conclusion of the Lunar Orbiter 3 Mission, the original LO task was essentially completed. Eight potential landing sites had been identified for the first Apollo missions. Cliff Nelson turned to Norm Crabill and asked, "What do we do now?"

Norm's answer, "Let's map the Moon.[1]"

Therefore, when Lunar Orbiter 4 was launched on May 4, 1967, it was destined to do just that.

In fact, LO 4 mapped 98 percent of the entire lunar surface. The front side photographs would be 70 to 100 meters resolution and the backside would be from 70 meters to 1.5 kilometers resolution.

35-LO 5

With the launch of LO 5 on August 2, 1967, we had launched all of the LOs in less than one year – to be precise, it was actually in 356 days. We had stayed on schedule with the exception of a two month slip in the launch of LO 1.

Lunar Orbiter 5 was designated as a mission to mop up the loose ends from all the other LOs. It was to provide:

- additional photography of the Apollo sites

- photographs of the areas not yet imaged by earlier LOs

- coverage of future surveyor sites

- coverage of landing areas for possible Apollo follow on programs.

These Apollo follow on programs were, of course, never to materialize.

One of the most interesting photographs we took on this mission was not of the Moon, but a photo of the near full Earth. India is visible in the center of the photograph with the Arabian Peninsula visible to the west.

This was a fitting end to the LO Missions: to look back at Earth one more time.

Photo of Near Full Earth
Taken on LO 5 Mission
Credit: NASA 1967 L 0677

36-LO Missions

We had our problems just like any space mission but we were able to work around them or minimize their effects as much as possible.

On missions 3 and 4, we lost a few photographs due to failures of the film advance motor that moved the film through the readout section of the camera system. Also, on mission 4 we saw from telemetry that the camera thermal door, which protected the lens from cooling too much and also kept out stray light, was not closing properly. We were afraid if we closed it all the way, we would not be able to get it open again. Therefore, we left it halfway open.

We had to make sure that light did not get into the camera lens and fog the film. At the same time, we had to assure the camera was looking near enough to the sun to keep the camera lens warm enough so there would be no condensation on the lens. This was done by carefully orienting the spacecraft to keep the proper sun angle on the camera side of the Lunar Orbiter. These types of problems and workarounds tend to be standard for space missions.

It is only with sharp and knowledgeable operations teams and with robust spacecraft that these space missions are as successful as they are.

We were nearing the end of the fifth and final mission of the LO Project. We had about one day to go before the final pictures were taken and then transmitted back to earth.

The maneuvers for the final photographs were loaded. Cliff was sitting beside me, and we were watching some of the last pictures come to life on the light board.

Cliff asked, "What are you going to do when you leave the SFOF after this mission?"

"Peggy and I are going to take a leisurely drive back across the country. There are still a lot of sights we haven't seen yet. I have no deadlines waiting for me, and this seems like a good time to do it. How about you, Cliff?"

"I am going to get into the boat and go sailing."

"Where?"

Cliff looked straight at me and said, "Does it make a damn?"

"No Cliff, it doesn't." And that's the way it was.

We all shared the pride that Lunar Orbiter had been an extremely successful project and had set the bar very high for those who followed. At the same time, we were all so worn out from the unrelenting stress of each of the missions and always preparing for the next one that any relaxation was a welcome respite.

It didn't really matter in what form it came.

37-Lunar Orbiter Retrospective

We launched five Lunar Orbiter missions in slightly less than one year. We slipped our initial launch date by two months but held to our schedule of launching an orbiter every three months from that time on.

After the fifth LO mission we had:

- Completed all Apollo photographic requirements in the first three missions by photographing all proposed landing sites.

- Found sufficiently large areas that met the Apollo criteria such that a safe landing could be planned.

- Used the last two missions to photograph the entire Moon.

- Proved that the radiation environment and micrometeoroid flux were sufficiently benign to allow the Apollo mission to move ahead as planned.

- Produced a better understanding of the lunar gravitational field. However, the job was not completed with LO, and not with Apollo. The determination of the lunar gravitational field is still an ongoing study.

Lunar Orbiter Badge
Issued after the Fifth LO

Not Bad For A Bunch Of Plumbers.

38-Lee Scherer's Coat

Possibly one of the little known factors that contributed to LO success was Lee Scherer's coat.

Lee was the Program Manager for LO, and on the first mission he came into the operations center wearing a plaid sport coat. He would come into the center, hang up the coat in the mission advisors area, and go about his business. The coat was present during the entire mission.

After the first successful mission, we all said that Lee had to wear that coat on all of the rest of the missions in order to bring us continued good luck. (He is wearing the coat in the picture of the press conference in which I talked about the Newcomb maneuver.)

We decided, after the fifth LO, we needed to do something with that coat as a memento of the successful LO flights.

After much discussion, we came to the only logical conclusion we thought was fair. We cut up Lee's coat and gave each member a piece.

Lee got the biggest piece. It seemed only fair, since it was his coat.

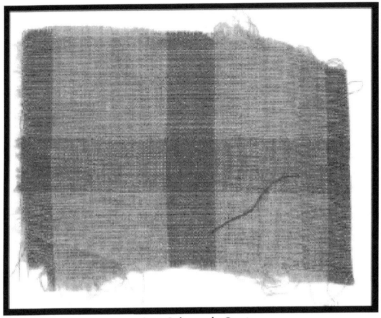

Lee Scherer's Coat
Courtesy of Paul Seamers
I lost mine!

39-Apollo Landings – One Giant Leap For Mankind

Now the results were in. Ranger had given us very high resolution photographs of the lunar surface as it continued taking pictures on its impact trajectory. Surveyor had landed in five different locations and given us a good understanding of the lunar soil. Lunar Orbiter had photographed all of the proposed landing sites in approximately one meter resolution and had provided additional information on the lunar gravitational field, the radiation, and micrometeoroid environment.

Now the final site selection for the landing of Apollo 11 could occur...and the winner was...Mare Tranquillitatis (Sea of Tranquility). The scientists had the full book of data on this one. They had the Lunar Orbiter photographs, the results from the Surveyor 3 that had landed in that area, and Ranger 8 that had impacted there.

As everyone remembers, the Apollo11 mission went well, and Neil Armstrong and Buzz Aldrin took their place in history when they performed the first manned landing on the surface of the Moon.

However, during the descent the old lunar gravitational field made itself known again. As Armstrong and Aldrin made their descent to the lunar surface, they realized that landmarks were traveling under them earlier than they should have, and they landed quite a bit down track from the planned landing point. This down track "error" was due to several error sources, one being Armstrong's maneuvering while attempting to find a landing site.

However, another significant error source was due to our old nemesis, the mascons and the rough gravitational field.[1]

While the nation rejoiced over the success of the program and the landing, the scientists told Apollo that they needed to have the landings in walking distance of the desired point, or the results would not be very useful.

Matt Grogan, who had led the navigation team for Boeing on Lunar Orbiter, had joined the Apollo navigation team at the Johnson Space Center in Houston soon after the last Lunar Orbiter mission. The Apollo navigation team, led by Emil Schiesser, had been working on a possible fix for the down track error problem for some time. Bob Tolson from LaRC had developed an additional mathematical term to help describe the lunar gravitational field in an attempt to increase the landing accuracy. [2]

The team finally developed a method to estimate a correction to the LEM landing site to compensate for the down track error. This correction was transmitted to the LEM, and the crew manually input the correction into the LEM guidance computer as they were descending to the lunar surface. A bit of a risky maneuver.

The first time these new systems were used was on Apollo 12, so when Pete Conrad and Alan Bean landed the Lunar Excursion Module (Intrepid) in Oceanus Procellarum, they could literally walk to the intended site which, in this case was Surveyor 2. They even brought home parts from the spacecraft.[3]

So Apollo was able to develop its own version of Kentucky Windage that worked for them, just as it had worked for Lunar Orbiter and for the old Kentucky marksman.

The Viking Days

40-The Viking Team

The creation of the Viking Project Office in December 1968 had been a long time coming. There was a whole lot of history leading up to it, starting seven years earlier. When NASA began looking at missions to Mars we had been busy developing and flying Lunar Orbiters. Many studies looked at different mission modes: soft landers, hard landers, orbiters, etc. The mission that finally began to take shape was called Voyager, and it consisted of Mars landers and orbiters. However, the cost continued to grow, even in the study phase, and in August 1967 it was finally canceled.

As Voyager faded away, Langley and JPL were beginning to study other missions to Mars. The launch vehicles had gotten stronger and better developed so more ambitious missions could be considered. Jim Martin was leading a team made up of members from the different centers to look at these options. The number of options was staggering. Over 20 options were being considered. During this time, from the end of Lunar Orbiter in 1967, all of Buglia's Boys, including me, were helping develop the different missions.

Jim Martin's team proposed two main options. The first option would have the Mars lander going directly to the surface of Mars without going into orbit first (direct entry option). A second vehicle would fly by and support the communication from the lander back to Earth.

The second option would require a larger booster, the Titan Centaur. In this option, the orbiter and lander would go into orbit around Mars. The lander would separate, make a soft landing on the surface of Mars from orbit (the out of orbit option), and the orbiter would continue to fly over to support the lander and continue to take pictures of Mars, measure the Mars surface temperature, and measure the water content in the Mars atmosphere.[1]

In thinking about these options, remember the world situation in the mid to late sixties. There was not just a Moon race, but there was also a Mars race with the Russians. Between 1960 and 1964, the Russians had launched six spacecraft toward Mars. All failed. We launched Mariner 3 and Mariner 4 in 1964. Mariner 3 failed, but Mariner 4 flew by Mars and successfully returned pictures of the Martian surface. We knew that Russia was planning four launches in 1971 and four launches in 1973. What we did not know at that time was that all eight missions would fail (making a total of fourteen failures), with only one of those spacecraft making it to the surface of Mars. We will talk about that mission later.

The two options (direct entry and out of orbit) were presented to Dr. Thomas Paine, Acting NASA Administrator, and Dr. John Naugle, Associate Administrator, NASA Office of Space Science and Applications, in mid 1968. Much of NASA and Langley management were in favor of the direct entry option, because this cost less, and they did not think Congress would look favorably on a higher cost mission.

Jim and the team heavily favored the out of orbit option because it presented a lower risk. We could look at the landing site and perform the final site selection before committing the lander. It also allowed us time for more analysis if we had a problem on board the orbiter or lander, since we could examine and hopefully solve the problem while in orbit about Mars.

In December 1968, we were happily surprised to hear that Dr. Payne and Dr. Naugle had selected the out of orbit option.

"Looking back," Naugle recalled: "It is a little hard to recapture the mood of the times...but...one of the things that figured in my mind was the fact that we were in competition with the Russians. They had a good strong program of landers, and I... felt that we had to establish a good, solid scientific mission. If the Russians landed successfully in '71 or '73, what we landed...had to be something that would stand up against what they had done."[2]

Naugle's and Payne's selection of the out or orbit mode, was a major factor in the success of Viking.

Langley Research Center, Hampton, VA, would manage the overall project and be responsible for the lander. JPL would be responsible for the orbiter and supplying the Deep Space Net and the Space Flight Operations facility from which we would control the mission. John Glenn Research Center, then Lewis Research Center, would be responsible for the Launch Vehicle. So the mission mode, the management structure, and the Viking name were decided.

Edgar (Ed) Cortright, who had become the Langley Center Director in May of 1968, and who would be a staunch supporter of the Viking Project, announced the formation of the Viking Project Office on December 6 of that same year. Most of the members of that core, shown on the next page, came from the Lunar Orbiter project.

Team Member	A Little History..
Jim Martin Project Manager	Had been Deputy to Cliff Nelson during Lunar Orbiter; would prove to be the absolute best man for the job
Israel Taback Deputy Project Manager and Engineering Manager	The best Chief Engineer I have ever worked with ; unbelievable problem solver and a guide to all of us. All of us were his students and grateful for his guidance
John Graham Missions Operations Manager	Led the video processing effort for Lunar Orbiter; then headed the Recovery Operations for Project Mercury
Gerry Soffen Project Scientist	An exobiology student (the study of the possibility of life elsewhere in the universe) when exobiology wasn't cool; proved he could handle a group of extremely intelligent and tough scientists
Norman Crabill Mission Analysis Manager	Played that same role in the Lunar Orbiter Project
Tom Young Science Integration Manager	Played several roles on the Lunar Orbiter project; continued in major roles on Viking
Gus Guastaferro Executive Engineer	Started managing projects with the Air Force; had been Launch Control Manager for the Scout missile firings from Vandenberg, CA.

41-Joining The Team

Now that the nucleus of the Viking Project Office had been formed, "Is" Taback and Norm Crabill met with me and asked me to come into the project office.

"Well, Newk, it is certainly your decision. I hate to lose you, but if this is what you want to do I certainly wish you all the luck in the world. And if you get disgusted, you will always have a home back here."

The speaker was "Bull." I had been working under him, even though I had been supporting the Lunar Orbiter Project, and later, the development of the Viking mission full time. This project, which would ultimately place two vehicles on the surface of Mars, would be much bigger and tougher than Lunar Orbiter, and would need a sizable number of people to manage the effort. This idea of leaving Jim Buglia's group was a more emotional decision than I had anticipated.

The term "Buglia's Boys" was in many ways real. We were like a bunch of closely knit college kids playing with mathematics and pulling crazy stunts, and Bull was trying to keep us in line, both academically and socially. One of Bull's favorite expressions when he found out something we had been doing was: "I'm surrounded by idiots." And I am sure that sometimes that was an accurate statement. I knew where I needed to go for the next chapter of my life, but still I hesitated. Finally "Is" said, "John, do I need to get Ed Cortright (Center Director) over here to ask you?"

I said "No, I'm coming." And the decision was made. I would be the Trajectory Design Manager under Norm Crabill in charge of designing the interplanetary, orbital, and descent trajectories. The most interesting part was doing those things while trying to meet the desires of the scientists, developing the trajectories, and determining the major requirements for the orbiter and lander. This would be Norm's ultimate responsibility, but the next level for accomplishing the work would be shared between Bob Tolson and me. Bob would be responsible for the Viking navigation. This entailed tracking the spacecraft with the three 200 foot antennas at Goldstone, California; Woomera, Australia; and Madrid, Spain; analyzing the data at the Space Flight Operations Facility (SFOF) at the Jet Propulsion Laboratory (JPL); and determining the spacecraft position and velocity and predicting where it would be in the future.

I began to build up my group with Otto[*], Burt Lightner, and Jim Youngblood. The project office also brought in General Electric as a support contractor, and several members of that contract began to support our mission design effort. Most of our efforts were now directed toward designing various missions that would develop requirements for the orbiter and the future lander. These lander requirements were to be used in the Request for Proposals (RFP) that would be used to select the lander contractor. This RFP to design and build the lander and to act as the integrator for the project went out in February 1969, two months after the official Viking Project Office was formed.[1] Obviously, we had been anticipating producing this RFP for a while, since you can't develop an RFP for such a major endeavor in that short a period of time. In reality, we had developed requirements for all kinds of landers, those that would go direct entry and those that

---

[*] Name changed to protect anonymity

would go out of orbit. Either of these options could land hard or soft. These approaches resulted from the mission studies I talked about earlier. Now it was decided that we would use the out of orbit, soft lander with an orbiter that would function for a long time in order to conduct its own investigations.

42-And The Winner Is The Martin Company

Now the pace, that was rapid already, sped up even more. The proposals came back three months later in April of 1969, and the evaluations began.[1] We were all cloistered in Fort Monroe in Hampton, Virginia for the evaluation. The three proposers were: The Boeing Company, Martin Marietta, and McDonell Douglas. After some evaluation, it became clear that the two lead proposals were from Boeing and Martin. Boeing's proposal was less expensive than Martin's, so that gave them an edge.

But Boeing seemed to have a problem. I was leading the Lander Trajectory Analysis Subpanel of the Source Evaluation Board. We began to question the volume of fuel in the Boeing lander deorbit system. It seemed to be too small. I phoned back to my guys at Langley and asked them to determine the probabilistic fuel requirements to deorbit based on the orbiter being in differing orbits. The answer came back. It wasn't enough. That was a severe problem for Boeing.

In order to make the deorbit propulsion tank larger and carry more fuel, Boeing would have to make the structure larger and possibly the aeroshell larger, which could require the parachute to be larger, which could cause the attitude control system to be larger, which could cause the propulsion tanks to be too small again, etc. This kind of system design analysis takes time, and you never know whether you have a system that works until the total design converges, and

Boeing did not have a system that would work as proposed.

On May 29, 1969, one month after proposals were received, it was announced that NASA would enter into a contract with Martin Marietta.

Since we had worked with Boeing on Lunar Orbiter, I had developed a real respect for the company, and I could not understand why they would make such a mistake. After the winner was announced and all the secrecy was over, I called Lowell Eldrenkamp of The Boeing Company. He had been my main counterpart during much of the Lunar Orbiter Mission design effort. I asked Lowell why they had proposed a design with that obvious flaw. He said, "John, the mission designers fought like hell against that design, but the managers wanted to keep the lander light in order to allow for future weight growth. We almost came to fistfights on the proposal development floor several times over that very issue."

Sometimes that's how it goes with the decisions that are made under the pressure of the proposal development.

With the decision made that Martin would be the contractor, the pace continued to increase. The detailed integration of the orbiter design with the Martin lander was being developed. This caused design changes in both vehicles, which then caused the capabilities and constraints associated with both of the vehicles to change. The science group was looking harder at the mission they wanted to perform. And we in the mission design group were trying to keep up. We had to take the science requirements, design the family of missions that would be necessary, and determine the major system requirements. We were at the front end of the requirements development process, which in this case was not a good place to be.

Also, the lander was going to fly into an atmosphere of unknown density and constituency to land on a surface of unknown altitude and bearing strength, with unknown slopes and rock sizes. Other than that we had it made. We developed a lander environmental system specification that attempted to encompass all of the possible permutations and combinations of these parameters. The lander had to be able to cope with anything inside this multidimensional envelope.

43-The Sol

As we developed the various mission designs, we needed a vocabulary so we could communicate. One thing that always required a bit of discussion was the simple term "day." Was the speaker talking about Earth days or Mars days, and if they were talking about Mars days what kind of Mars day were they talking about since there were different ways of measuring days? The duration of our Earth day is 24 hours long and is defined by the average time between two successive crossings of the Sun across a given longitude line. Astronomers call this a mean solar Earth day. We decided to use the same type of definition for a day on Mars, which would be the average time between two successive crossings of the Sun across a given Mars line of longitude.

The accurate description of this day was a Mars mean solar day, which is approximately 24 hours and 39 minutes. So just as a way to communicate, I said, "Let's call the Mars mean solar day that we will be using a sol ." And that name stuck. When we spoke of a day on Mars or a Mars day we would simply say a sol. I still have a small sense of pride now when I hear the TV announcer comment that scientists call the Mars day a sol.

44-Congressional Holdup

During this time we had continued to hear murmurings that Congress might cut our funding, which would mean our launch would be moved from 1973 to 1975. Norm took a quick look at that launch opportunity and concluded we could not get to Mars, because the launch energy requirements in 1975 were too great. He immediately told Jim.

Jim's response was very predictable, "Norman, you do not understand the problem! If we have to slip, we will launch in 1975, and we will get to Mars! Is that clear? Now go figure out how we will do it!"

There is only one answer to a statement like that, especially to Jim, and Norm said it: "Yes, sir!" So we took another look at the launch opportunity.

We had always designed missions based on Type I trajectories. Type I trajectories are those in which the spacecraft travels less than 180 degrees around the sun as it goes from Earth to Mars. It turned out that we could use Type II trajectories (traveling greater than 180 degrees around the sun) for the 1975 opportunity. This was the first time anyone, including NASA, had ever used these trajectories, so it was a new idea. We had not thought about it before, but necessity breeds invention, especially when Jim required it.

The murmurings of funding cuts quickly proved to be real. The Martin Company had only been under contract for eight months when in January 1970, Congress reduced our funding and caused the mission to slip from a 1973 launch to a 1975 launch.[1] Most of the project office troops were very worried. I heard Ed Brummer (the Spacecraft Manager) say, "This is usually the first step in canceling the project." This was the attitude of many of the project personnel.

I was elated. I thought, "By God, now we can do it right!" All of us had been working ten to twelve hour days, and six or sometimes seven days a week for at least a year, and we were in the first phases of the project. We had not even gotten to PDR (Preliminary Design Review). What was going to happen when we got to CDR (Critical Design Review) and then the part of the project where we complete the building and testing of all the individual subsystems and integrate and test all the subsystems as a system? And some of the subsystems are not ready, and things go wrong.

But the euphoria of being able to slip two years lasted for about fourteen minutes, until Jim Martin called Norm, Bob, and me into his office.

He started off by saying, "You know we have been slipped from a 1973 to a 1975 launch."

"Why yes, we know that."

"And you know that I have issued a stop work order on Martin and JPL."

"We know that."

"And you know that the orbiter and lander are waiting for you to develop a new requirements document for the new launch opportunity. They need this document before they can begin developing the orbiter and lander in detail. How long will it take?"

We were ready for that one. We could see that coming a mile away under bright lights. We had developed the same set of requirements for the 1973 launch opportunity over about a year as we studied the different options. So we felt that we could develop the new set in a slightly shorter time.

"We can do that in nine months," we replied.

"You do not have nine months."

Then we started negotiating, Jim Martin style, meaning we ended up doing what he told us to do. We agreed to get a 90 percent answer in three months and a 100 percent answer in six months, and we were told that no one expected to see any real changes after the first three months.

The pace stepped up again. It was during these next six months that I lost two good men, both due to the stress of the Viking project.

The first was Brad.[*] Brad worked on Viking at the Martin Company, and was an internationally known figure in guidance and error analysis. He was also a strong individual, but sometimes that doesn't matter. We were working to redevelop our new requirements document when I got a call from Brad's wife. She said Brad had come home from the office in the middle of the day and was emotionally unstable, and it was due to the stress of the project. (In reflecting on this, it shows how closely we worked with our Martin contractors, that I would get a call from one of their wives to ask for help.) She wanted me to help convince Brad to check into a hospital, which I did, and we were successful. Then about 4 weeks later I got a call from his doctor.

---

[*] Name changed to protect anonymity

"John, Brad is feeling much better and wants to go back to Viking. What do you think?"

After I asked some questions to try to get some feeling as to Brad's emotional strength, I said "Not only NO, but HELL NO! Viking is a pressure cooker, and there is no way he can go back to Viking and operate at half pressure. Also, there is no way I can protect him from 2000 miles away, and there is no way his boss can protect him. From what you have told me, he is not strong enough to take the pressure he will be subjected to."

"Well, I will see what I can do, but Brad really wants to continue Viking."

"I have told you, and I will tell Brad if you want me to, but I think it is a big mistake."

Two weeks later Brad was back on the job.

Four weeks later, I got the call that Brad had a bad day at the office, went home, started the car, and put the hose in the window. He had gone to sleep. Permanently.

The next casualty was Otto*. Otto worked directly for me, and was a good analyst. When we realized that we were going to have a smaller launch period in 1975 than we did in 1973, I suggested that we look at a different type of trajectory that might open up the launch period. In these trajectories the spacecraft would have to make a large velocity maneuver about half way to Mars to change its plane of rotation about the sun, hence the name broken plane trajectories. The problem with these types of trajectories was that the mathematics associated with them had never been developed. This required the development of the mathematical algorithms and a computer program that would compute the trajectories and search for the minimum velocity solution on any given day.

I told Jim Martin about this idea and that I needed money to get this package developed so we could examine this type of trajectory.

Jim's answer, "Do it, and let me know what it will cost. And realize that the algorithms and software have to be developed and analysis completed in six weeks. That is when we have the next Viking Management Council meeting. If we are going to use this, we need to know by then."

"Okay – no pressure!"

The Martin Company could not attempt it. I knew the mission design leaders at many of the big corporations. Ours was a small and somewhat closely knit community. So I gave each of them a call, explained the problem, and then let them know the deadline. Everyone's answer was, "John, there is no way I would sign up to that contract. I don't care how much you pay me. I am not sure it can even be done and certainly not within your time limit!"

Well, that didn't work. So I met with Peter DeVries, Frank Nicholson, two of our GE support contractors, and Otto, and we started to define a way to attack the problem. At the next weekly project meeting, I told Jim that no one else would commit to performing the analysis, so we would do it. I put Otto in charge of the effort. For the next few weeks, we had daily ten minute tag ups in the morning to assess the progress. Finally, just in time for the Management Council Meeting to be held in Denver, the job was completed, and the analysis performed. Otto was to give the presentation. This was a trajectory type that had never been flown, and therefore was risky. So we knew that we would get hammered with questions, rightfully so. But we had prepared Otto. He would be going through the presentation to the Management Council, someone would ask a question, Otto would look at his slide list, and say, "Give me backup 14, please." He had as many charts in the back up pile as he had in the actual presentation.

However, the broken plane trajectory idea never got off the ground. The navigators, the ones who had to track the spacecraft and define the trajectory, were extremely worried that they could not define the trajectory after such a large velocity maneuver that far away from earth. Also, the broken plane concept added more days to the launch period at the beginning of the period. If it had added more days at the end of the launch period, we might have considered it, even though it was risky. As soon as I realized that the concept would add days at the front of the period and not the end, I knew we would not use it. I felt the project made the right decision not to consider the broken plane concept.

However, the day ended on a good note. We all went out and got drunk, making sure that Otto got his drinks for free. I do not remember the name of the hotel bar into which we ambled. We were going to stop for a drink and then go to dinner. We never made it to dinner. They served "bull's horns," which was some type of drink I have never had before or since, in a ceramic container that looked like, guess what, a bull's horn. I tried to smuggle one out under my jacket but the waitress caught me. It was pretty obvious, since the bull's horn was sticking out of the top of my jacket. I always have wished I had bought the damn thing. I would have loved it for a souvenir.

We came home from the Denver trip on a Friday. About halfway through the following Monday, Otto came into my office.

"I've got to go home," he told me.

He looked like a wild man. His face was distorted. Was it anger? Fear?

"What is the matter, Otto?"

"I just have to go home. That is all."

"Of course, Otto."

And off he went. I sat at the desk for a long time trying to figure out what I had just seen. Not having a clue, I decided I needed to talk to Otto some more. I went into Norm's office and explained what I had just seen.

"I don't know what's going on with Otto but it isn't good. I am going over to his house."

Norm nodded. "Go, and let me know what you find out."

Otto let me in, still with that unexplainable look on his face. "You got an extra martini, Otto?" I knew that was his favorite drink, and I figured he would have one with me. After talking about the Denver trip for a while, congratulating him again on his performance, and waiting for the martini to become absorbed, I asked the real question.

"Otto, is anything wrong?"

"You're damn right there is! They've tapped my phone! And I think they're spying on me! I think it's my wife, because she is running around on me. She doesn't want me to find out."

The conversation went on this way, through at least one more martini. When I realized I was getting the same information over and over, I excused myself and headed home. It only took a five minute discussion with my wife, Peggy, to nail it.

"John, you're looking at a textbook case of a mental breakdown."

And that turned out to be the correct answer. So I was back working with one of my guy's wives in order to get Otto some help. We did get help for him, and he eventually came back to work and was productive for a while. But toward the end of the development phase of Viking, the pressure began to take its toll again. Finally Otto took a medical discharge, and that was the last I saw of him. Another good guy crushed under the pressure of Viking.

45-Landing Site Selection

The new Requirements Document had now been developed. Gerry Soffen, Project Scientist, and Tom Young had put together the Viking Science Team. The team was very impressive and included Nobel Laureates, such as Dr. Joshua Lederberg and other well known scientists, such as Carl Sagan. It was during this time that I got to witness the skill with which Gerry could handle these very intelligent scientists that had egos to match.

This was the era when Henry Kissinger was going all over the world performing what the press labeled "shuttle diplomacy." Gerry had a very pronounced shuffling step. When there was a large disagreement in the science team, I would watch Gerry shuffle over to one of the team members, talk softly, and then shuffle to another member and again talk softly with him. Pretty soon, as if by magic, the problem would diminish or go away entirely. I came to call what Gerry did as "shuffle diplomacy." It was beautiful to watch, and whatever it was, it worked.

The smaller landing site working group was formed in September 1970.[1] It was made up of members of the Science Team with Tom Young as chairman and Gerry Soffen as the Project Scientist at Tom's right hand. Howard Robins had joined the project and stood in for Tom for a year when he was chosen for a Sloan Fellowship at MIT. The team also included the principal investigators and team members of the orbiter and lander experiments. The debate over the most appropriate sites would continue for almost three years until May 1973, when the final selection was announced.[2]

As the Landing Site Working Group began its deliberations in 1970, there was very little data to work with. Mariner 4 had flown by Mars in July 1965 and had taken about 20 pictures, with poor resolution.[3] The pictures that did come back showed a Mars that looked like the Moon. It looked like a barren, primordial planet. Mariner 6 and 7, launched in February and March 1969, flew by Mars in July/August of that year and took about 60 near encounter photos. In all, the photos of Mariner 6 and 7 covered about 20 percent of the Mars surface at varying resolutions.

Mariner 9 was launched in May 1971 and went into orbit about Mars in November of that year. With the Mariner 9 imaging we began to see a different planet. But we did not see these things right away. As the spacecraft approached Mars, the astronomers watched as what appeared to be a large, white cloud engulfed the planet.

This was the largest, most expansive dust storm ever recorded on Mars. We had to wait in orbit for a few months until the storm dissipated in order to get clearer pictures.[4]

However, the Russians were not that lucky. They could not wait. They had launched 2 orbiter/lander spacecraft (Mars 2 and 3) that same year, and they arrived in the middle of the storm. Their two landers separated from the orbiters as they approached Mars and made a direct entry into the Mars dust storm.

One lander, Mars 2, crashed on the surface of Mars. Mars 3 made it to the surface and relayed 20 seconds of information. I use the term "information" loosely. "Mars 3" sent back 20 seconds of the Russian National Anthem![5] Obviously, this was not a scientific decision. We were disappointed the Russians could not tell us anything about the atmosphere or the Mars surface. But when we saw the Russian scientists, they were about two feet off the ground because they could have gotten some very important data in those 20 seconds!

But they got nothing.

The two Russian Mars orbiters did not last long enough to see through the storm. It was still raging when they died.

As the veil of the dust storm gradually lifted, Mariner 9 sent back photos, and we were able to see features we had not seen before: huge canyons, volcanoes, river beds, uplifts, and fault zones.[6] But we were still looking at Mars with a resolution of 1000 meters through Mariner's cameras.

You could have thrown down a bunch of battleships and an aircraft carrier or two on the surface of Mars and not have been able to see them with the resolution of the Mariner cameras. But this was the best data we had, and the site selection continued based on this information.

46-Go To The Water

During this time many different sites were chosen. We developed possible missions to all of these sites, continuing to examine the requirements they placed on the orbiters and landers. The question of the search for evidence of life played a large part in those discussions. When that question came up, the discussion immediately turned to water and where it might be. Dr. Joshua Lederberg (a Nobel Laureate in Genetics) began to strongly advocate going to 65 degrees north near the edge of the polar cap.[1] That sounded like a good idea except for the fact that Viking was designed to land between +/-30 degrees latitude. This opened up the discussion of going out of the +/-30 degree box and caused a lot of angst in all of the Viking engineers. This discussion lasted for several months, as different ideas and sites were considered. Several Viking engineers, including Norm Crabill, talked with Dr. Lederberg and tried to explain the problems with going that far north. It was during this time that Jim Martin called me into his office. "John, I want you to go out to California and explain to Dr. Lederberg that Viking is designed for +/-30 degrees and that we cannot go to those higher latitudes. You have just got to make him understand the problems associated with those higher latitudes. "

"Sure Jim, I can handle it."

Now I would learn how much power Nobel Laureates commanded.

They packed the lander specs and design information under

my arm and sent me off to California. I spent the best part of a day with Dr. Lederberg, explaining why we could not do what he was proposing. I thought I did a pretty good job.

At the end of the day, Lederberg sat back in his chair, looked at me, and said, "John, I have enjoyed your visit. I have learned much about the lander that I didn't know, and I appreciate that. However, I want you to take this message back to James Martin:

- The purpose of Viking is to find evidence of life on Mars.

- The best place to look for that life is where there is water.

- The most probable place for water is at higher latitudes at approximately 65 degrees.

Therefore, if Viking does not go to these higher latitudes, I will simply write the President and tell him to cancel the program."

Well, that seemed pretty clear. I carried that message back to Jim, and we moved on from there.

As the discussions progressed, we began to consider the idea that the more northerly latitudes would obviously be colder and not as hospitable to life. The site selection team began to consider moving closer to the equator than 65 degrees north. Finally, a compromise was reached. The first lander would go to approximately 20 degrees north. If the first lander was successful, the second lander would go to approximately 45 degrees north. If the first lander was not successful, the second lander would go to approximately 5.5 degrees south, which was deemed to be as safe an area as we could find.

These sites were announced by NASA in May 1973.[2] With these final preselected sites we moved out with the final mission designs.

47-Mission Operations Design

Not long after the sites were selected and the mission designs had been somewhat filled out, Jim asked me to move over and support the Mission Operations Manager, John Graham. John had led the development of the Video Processing Laboratory for Lunar Orbiter and had also been the head of recovery operations for Project Mercury. You probably saw him on TV. John was always on one of the aircraft carriers that received the capsule after it had been lifted out of the ocean. When Gordon Cooper's capsule landed on the deck of the aircraft carrier KEARSARGE, John was there to welcome him home and help him out of the capsule.

John now had two staff men – Curt[*], who had been hired from another center, and me.

As I came into Mission Operations, John presented me with many tasks that needed to be completed within a short time period and that had to be reported at the next Mission Operations Review Board. Some of these tasks were assigned to me and some had already been assigned to Curt. I thought, "Here we go again. Too much work and way too little time". The pressure was on already! And Curt was not getting his job done at all! I was walking beside John as we came out of the first monthly project review since I had been assigned to Missions Operations

I said, "John, that guy is not cutting it."

---

[*] Name changed to protect anonymity

"I know. I'm going to fire him."

I said, "Okay," and I went back to my office. I had just gotten into my office when John called.

"John, I've made a decision. Come to my office, now!"

As I walked in, John said, "I was serious. I am going to fire him. And you are going to have to pick up his load!"

"Okay, but you will have to give me some more people!"

"I can't give you any more people!"

"Well then, damn it! Give me some more time!"

"I can't give you any more time! I can give you Curt!"

"I don't want Curt! He got us into this mess in the first place!"

"Well, give him an interview and see if you can use him! He is the only other person I have!"

"Alright. I'll give him 30 minutes! That's all this decision is worth! Send him up to my office!"

By this time, we were each standing on opposite sides of a small conference table, nose to nose, screaming at each other. I stormed out of John's office to face a small group of snoopy listeners with grins on their faces. They had been enjoying the "discussion."

Someone had the audacity to say, "Bad day at the office?"

I immediately told him where to go and exactly how to get there.

John sent Curt up to my office for the interview. After about

15 minutes I decided this wasn't going anywhere,

I suggested that he leave. I picked up the phone, dialed John's number, and said, "John, this is John. Fire him!"

I never saw Curt after that. In retelling this story a few years later, I realized that I never even asked about him. Things were moving so fast and furiously that we were immediately forced to focus on other problems, and what happened a few minutes ago was history unless it affected what was going on at the moment.

My main job under John Graham was to be the Operations Analysis Manager and lead the development of the Mission Operations Strategy. This was the complete protocol we would use to fly the Viking mission. We would determine what needed to be developed, what groups of the flight team would develop it, and what they needed to develop it. We would also decide who had the authority to allow the product of one group to pass to the next group. This process would span the flight team efforts from the time the scientists said, "I think we should do something," until we had commands coming out of Goldstone, California or another Deep Space Station. We also had to close the loop by developing the same type of protocol for the down link portion of the mission. This protocol had to be robust enough to redesign the mission in real time. We would use this feature of mission redesign continuously during the site selection phase of the mission as we tried to find acceptable sites for the Viking landers. We did not know that phase was going to last for 76 days.

As I began to develop the Mission Operations Strategy (MOS) there were many items that needed to be nailed down quickly. So I was off on multiple trips, primarily to JPL and The Martin Company. On some of these trips I would leave the house, take an early flight from Norfolk, VA and fly into the Los Angeles Airport (LAX). Then I would get cleared through security, climb the stairs to the roof, and meet the helicopter that had been sent to carry me to JPL. We would fly to JPL, hold the meeting, and the helicopter would fly me back to the roof at LAX so I could catch my plane home. It was like being an actor in a James Bond movie.

The plane home was the red eye and left LAX at midnight. The first time I did this I thought I would be able to sleep all the way across country and then go into the office reasonably well rested the next morning. But then I realized this flight was the second leg of the Royal Hawaiian, and these people had been partying since they left Honolulu. So much for sleep. I finally decided that I might as well join them, which I continued to do on subsequent trips. The following day at the office was a rough one.

In order to complete the development of the Mission Operations Strategy (MOS), I brought all of the heads and prime movers of the different flight team groups to Langley. I told them to get an open return ticket because we were not going to leave until we had this process nailed down. They came ready for work, and within a week we had the core process completed.

Now it was time to present the MOS to the Viking Management Council. We looked at several ways to show the MOS to the Council, and we finally decided to draw a flow diagram of all of the activities that would be going on from the time of the first Viking's arrival at Mars until after the second landing. We started describing the operations with two spacecraft approaching Mars, and then with the first going into Mars orbit. We continued describing the operations with three spacecraft after Lander 1 was on the Mars surface and then with four spacecraft after the second lander was down. The chart that was generated to represent all of this activity was about 20 feet long and 4 feet wide.

I was first on the Management Council agenda. The presentation was supposed to last an hour. After the presentation started, it was clear that we were going to run over the allotted time. Jim's comment was, "Scratch all the remaining items off the agenda. We are going to work this until we understand it."

And so we did. That was the longest single presentation I have ever given in a project environment. It lasted for six hours.

When we broke for lunch, I walked out of the conference room beside Jim. He looked down from his 6' 4" height and said, "Good job." Then he looked at me again, shook his head, and went "ugh." Jim was famous for his "ughs." It meant something was bothering him, and he wasn't sure what it was yet. By the end of the meeting, Jim knew what the problem was and how to fix it.

We originally had the two spacecraft arriving at Mars thirty days apart. Jim now realized how much work had to be done by all of the different members of the flight team in very short periods of time. With the team working so fast and furiously, he was sure someone would make a mistake. His final decision was to pull the arrivals of the two spacecraft apart by fifty days instead of the original thirty.

It turned out we would need every minute of those fifty days.

48-The Orbiter And Lander Come To Life

While the missions were being developed and the sites selected, the hardware guys were busy building an orbiter and lander.

The orbiter and lander designs were shaping up very well. The orbiter would carry the lander to Mars, enter Mars orbit, and act as the platform from which the landing sites would be observed. The lander, which was encapsulated inside the entry vehicle, would then separate from the orbiter, and after accomplishing its entry through the Mars atmosphere, would land on the Mars surface.

The total configuration would look like the picture on the following page. Shown at the bottom is the lander encapsulated in its bioshield, which consisted of the base cover and cap. The top structure is the orbiter. Some of the major orbiter structures that are visible are the solar panels, the high gain antenna, and the engine nozzle for the orbiter velocity maneuvers.

The Viking Spacecraft
Orbiter above with the Lander folded up
and encapsulated below.
Credit: NASA

This picture, taken in the Midwest, shows the lander as it would appear on Mars, but obviously without Carl Sagan standing beside it.

Credit: JPL NASA

49-Stuff It In A Box

With any spacecraft development, there is always the problem of how to get everything into the available space, and we had those problems. One of the worst was the biology experiment. This consisted of three different experiments; each designed to detect life or evidence of life on the Mars surface. Also, there was a gas chromatograph mass spectrometer (GCMS) to look for evidence of organics. Normally, this set of hardware (the biology instruments) would take up a medium sized laboratory. Now that entire apparatus would have to fit into a cube that was about a foot on each side. This caused the Biology Instrument Manager, to lose his hair, literally, and caused several letters to be written from Ed Cortright, who was the Langley Center Director, to the Executive Vice President of TRW (the company building the instrument).

The instrument was finally completed, successfully tested, and delivered to the lander for integration. I have included a diagram to give you an idea of how tightly packed our finished product looked.

An unbelievable accomplishment.

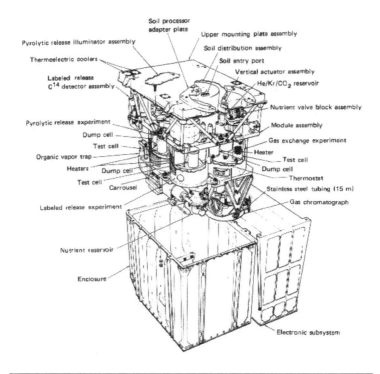

Biology Instrument Design
Credit: NASA

50-We Are Coming In Fast

This was the first time we had traveled through an atmosphere other than Earth's, which was a major challenge. The entry capsule that separated from the orbiter consisted of an aeroshell and base cover that was designed to protect the lander from the high temperatures expected during entry. In addition, the entry capsule contained a parachute system for additional deceleration. Each of these systems was specifically designed for the Viking mission. The entry system and trajectory design was complicated by the uncertainly of the composition and density of the Mars atmosphere. When Mariner 69 flew by Mars, the results of the occultation experiment showed the atmospheric pressure at the Mars surface to be between 4 and 20 millibars. This is equivalent to between .06 psi and .3 psi. Our atmospheric pressure here on Earth is approximately 14.7 psi at sea level. The Mars surface pressure is equivalent to the pressure in our atmosphere slightly over 20 miles up, where an astronaut would have to wear a space suit to survive. This is also close to the highest altitude a jet aircraft has ever flown.

The original design of the entry system assumed the atmospheric pressure at the surface of Mars was around 20 millibars or .3 psi. Now we realized that the atmosphere could be much thinner than we originally thought. The aeroshell that would slow down the lander as it first entered the atmosphere was already at its maximum size. We were already planning on flying the lander so it would generate lift and therefore stay in the Mars atmosphere longer to allow more time for decelerating. The only remaining thing we could do was to make the parachute larger. This added more weight, which we did not need. In addition, the parachute would have to be deployed at a speed greater than the speed of sound in the Mars atmosphere, which is approximately 540 miles per hour. (Speed of sound at sea level on Earth is approximately 770 miles per hour.) This was another first: deploying a parachute behind a blunt body at supersonic speeds. We blew out a few parachute models in the wind tunnels and did serious damage to a full scale model over the White Sands Test Range of New Mexico before we got it right. However, the studies, experiments, analyses and tests conducted to develop the Viking entry descent and landing system are still the textbooks for any new similar activity today.[1]

51-The Final Thrust

The final phase of the entry and landing sequence was accomplished by the terminal descent rocket engines. There was a problem with their design. Originally each of the three engines had a single nozzle with an opening at the end of each nozzle of approximately five inches. There was a lot of concern about the plume from the nozzles digging holes in the Martian soil and the lander careening into the hole or building up a mound of soil that the lander footpad would land on and cause the lander to tip over.

Another concern was the possibility of blowing away all of the Martian soil in the area and leaving only solid rock. In addition, we were concerned about blowing away the soil under the nozzles, resulting in the lander base plate sitting on the ground; this would cause a thermal problem. Also, we were worried about the possibility of destroying any of the organics our search intended to find. Finally, after much analysis and testing, it was decided to have 18 different nozzles on each of the three terminal descent engines, each with an opening of about an inch. This way we had 18 smaller, shorter, plumes rather than one big plume from each of the descent engines. The smaller plumes would have a very small effect on the soil.

Viking Terminal Descent Engine with 18 Nozzles
Image courtesy of Aerojet from Tom Dahl

52-Something Has To Control This Thing

It has been said, "The lander computer was one of the greatest technical challenges of Viking."

Remember, this as the late '60s. There were few computers in existence, and there were none that would do what we needed for Viking. We had constraints on weight, size, and power, and we needed to have a sufficient memory capability to store and execute the necessary commands. We experimented with different types of computers with little or no success.

Time was running out. The computer concept that looked most promising was the computer made by Honeywell. This computer was made with what was termed plated wire memory. This was constructed by covering the wire in the computer core with a magnetic coating so information could be stored on different sectors of the wire. However, the coating always had dead spots in it that would equate to a bad spot of memory in the computer.

Finally, it was found that GE had done some classified work for the Air Force and had solved the problem by using a special process of plating the wire with several coatings. This led to an entourage from Langley, including Fill Cuddihy, electronic systems; Ed Cortright, Center Director; Jim Martin; "Is" Taback plus several of the Martin people going to GE and getting agreement to use the GE technique. This finally enabled the completion of the computer so that it could come together with the rest of the subsystems during integration.[1]

53-Let's Cook It

One of the requirements on Viking that was very difficult to meet involved sterilization. There was an international agreement (primarily involving the US and Russia) that neither of us would contaminate Mars, meaning neither the US nor Russia would inadvertently take any Earthly biota to Mars. Since we were looking for signs of life, this was very important to Viking itself. We did not want to take biota to Mars and then get evidence of life, only to be unsure as to whether or not we brought it with us. This concern had been voiced by several scientists, including Dr. Joshua Lederberg.[1]

This sterilization requirement is also the reason that the Bionetics Corporation is now headquartered in Hampton, VA. We had sent out the RFP to select the contractor that would be in charge of examining methods for, and actually performing the sterilization efforts for Viking. One of the proposers was Bionetics, a small company headquartered in California, owned and run by Dr. Joe Stern.

In the questioning session that was part of the proposal evaluation, Jim Martin made the comment to Joe: "You know this sterilization effort is going to be rather complicated and will influence greatly how we build the lander. I would like to have the company that is overseeing this effort close to the project office. Would you have any objection to moving your company to Virginia?"

Joe, always a pragmatist, did not bat an eye.

"We would be glad to come to Virginia and set up our headquarters next to Langley so we could be available to the Viking project at all times." Then he went home and told his wife, "Guess where we are going to be living, Honey?"

We looked at different ways to answer the sterilization requirement and finally decided to cook the lander, which is what we did. After the lander arrived at the Cape, we conducted a thorough checkout of all systems and then encapsulated the lander in a bioshield base and cap (drawing shown). This meant that nothing, or no one, could come in contact with the lander after encapsulation. We had already constructed an oven that measured 30 feet by 30 feet by 30 feet into which the capsule was placed. We then baked the lander in its capsule for 40 hours at 120 degrees C (248 degrees Fahrenheit). We made one final checkout, mated the encapsulated lander with the orbiter, and tested the lander subsystems one more time.

If you do not think this sterilization was a tough requirement to meet, try baking any one of your electronic devices, like your cell phone, then using it. And remember, we were operating with late 1960's technology.

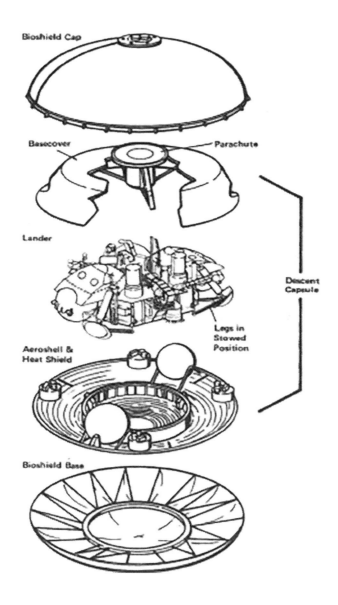

Viking Lander
Schematic Drawing showing Major Components
Credit: Reference 10

54-Test As You Fly

As the lander was being completed, the question of the final testing of the lander arose. The plans called for an end to end systems test. This is the test in which the lander would be placed in an entry condition and fed the same type of information that it would see during final descent and landing, in a complete simulation of those phases of the mission. The lander would then respond just as if it were executing in real time. This is the type of test we loved to perform, because it would tell us if all of the subsystems were communicating and working properly together, assuring us that the total system would work as it was supposed to when we reached Mars. This type of test would have caught the Hubble mirror problem, for example. Just as in all projects, this test came, necessarily, near the end of the system development, when the time and money are usually running out. This was the problem with Viking.

The arguments against this type of test are always the same. The subsystem guys say that they have tested the respective interfaces over and over, and that their subsystems have been tested and are performing their functions properly. Also, the system guys say that you cannot perform a perfect simulation, because you cannot simulate all of the conditions simultaneously; therefore, the test is not totally accurate. Also, Jim was taking a huge amount of heat from headquarters to trim cost.

Now we convened the final meeting to make the decision as

to what to do about the end to end test. All of the arguments were presented. "Is" Taback was adamant that we should conduct the test. His mantra had always been, "A test is worth a thousand opinions." I have heard him say that about a thousand times, and he repeated it now. "Is" and Jim, who had been hand In glove on all decisions through the Viking development, now saw things differently. This was the only time I saw "Is" lay his body across the track and Jim run over him. Jim's final statement was that we would not conduct the end to end test on Earth. That test would be conducted after we arrived at Mars as the lander made all of its required maneuvers and descended to the surface of Mars.

I made the mistake of walking out of the meeting beside "Is." He was furious.

"You know we canceled the test," he said.

"I know, I was in there, too."

"Is" then figuratively picked me up by the collar and said, "As of now I hold you personally responsible to ensure we do not have a sign reversal in the control logic."

"Is" was worried that, inside the lander computer, nestled somewhere in the thousands of values or functions, there might be a small value or function that had the wrong sign (positive or negative), and one wrong sign, out of thousands of signs, could kill us. The end to end test would hopefully find this error, if there was one. NASA had made that mistake before. And the mistake can come from many sources.

I thought, "Oh hell, there goes the next few months." I immediately ran down to the office of Milt Holt, who was managing the computer, and we devised a way to examine the problem. It turned out to be relatively easy, since Milt, as

I had expected, had been on top of this already. But it served as my panic for the day.

55-Playing With Firecrackers

Finally, after seven years of design, development, and testing... after all the problems...the landers and the orbiters came together at the Cape as integral systems; all systems worked, and they worked together. Each system was thoroughly checked out, the landers were encapsulated and sterilized, and the individual landers and orbiters were mated and made ready for the long trip to Mars.

Viking Orbiter and Lander
Credit: NASA

As always, we were fighting time. Burt Lightner, who had worked for me during the early days of the mission design effort, was now at the Cape. He had devised a complex launch date/arrival date strategy for the launches of the two Viking spacecraft. Using this strategy, the launch period opened on August 11, 1975.[1] We had a total of thirty days during which we could launch the two spacecraft and perform our nominal mission. If we went beyond the thirty day launch period, some of our mission objectives would have to be compromised.[2] It was actually even more constrained than that. In order to perform our nominal mission, Viking 1 had to be launched by August 20 to maintain the fifty-day separation time between the two spacecraft arriving at Mars.[3]

Now on August 11, 1975, Viking 1 and the Titan Centaur were ready to launch, we thought. But at launch minus 115 minutes, the Launch Vehicle Manager informed the rest of the Viking Launch Team that one of the twenty four thrust vector control valves on the Titan Centaur did not operate properly during tests. Those valves provided steering control to the Titan during the first boost phase of the flight. This caused the launch team to scrub the launch while the valve was repaired. The launch team did not know it then, but this was the beginning of one of the most stressful 30 day launch periods in history.

On August 13 the team met for a briefing on the launch vehicle progress and found that the orbiter's battery was extremely low. After more analysis it was believed that a lightning strike in the distance had inadvertently turned on the orbiter and caused the drainage from the battery. No matter the reason, the orbiter had a problem.[4]

Jim decided to remove the first Viking spacecraft and replace it with the second spacecraft that had been readied for flight.

On August 20, Viking 1 was again ready for launch.[5] As always, there were continuous weather briefings at the Cape. At the 10:30 a.m. briefing, the team was told there were thunderstorms in the area and a squall line off the coast. The thunderstorms would possibly begin at 2:00 p.m. and last for five hours. The count continued. At the 3:00 p.m. weather briefing the storms had not arrived, and the team was told the squall line was breaking up. It looked like the thunderstorms would hold off. There were sighs of constrained relief. The spacecraft and launch vehicle continued to check out properly, so at 5:22 p.m., August 20, Viking 1 headed for Mars on the very last day of its launch period.[6]

With Viking 1 on its way, the team turned to the launch of Viking 2. The schedule for the readying of the launch vehicle and the spacecraft was intense. Every hour of every 24 hour day was scheduled. And the launch drama became even more intense. After re-readying the launch facilities, the Titan-Centaur and Viking 2 headed for the launch pad on August 26 with fourteen days remaining in the launch period.

The final tests and checks began. On August 30 during these checks, it was found that the orbiter radio system was not working properly. The next day, after more analysis, it was determined that the spacecraft needed to be taken off the launch pad and the S-band hardware replaced. This meant the spacecraft had to be separated from the launch vehicle, returned to the assembly area, the subsystem replaced, and then rejoined with the launch vehicle. This is a complicated process, coupled with all of the tests and safety measures. This would cause the second launch to be on September 11, two days after the end of the nominal launch period, compromising some of the mission objectives.

This was the only time I have ever been aware of Jim Martin being openly discouraged. Don Ward, who was Deputy Project Manager for Launch Operations, saw Jim as the team was taking the spacecraft back to the assembly area.

Jim had worked the schedule on the back of an envelope. "We'll never make it! I don't see how we will get off before September 11!"

Don had worked the schedule as well." Jim, I think we can do it. Let me work it."

Jim's response, "You'll have to show me."[7]

Don assembled the Launch Operations Team, and the schedule tightening began. Otis Childress, who was Manager of the orbiter, working with JPL, contributed four hours out of their orbiter rework schedule. The launch operations cut the pre count schedule from twelve to ten hours. The spacecraft closeout time was cut from 8 to 2 hours, based on the fact that they had done it several times by now. The process continued.

The new schedule showed that they would miss the September 9 launch by twelve hours. They went after the schedule again. Otis and the orbiter team contributed another six hours, and the launch vehicle team contributed the remaining six. The new, compressed schedule worked, and Viking 2 was ready to launch on September 9, the last day of the launch period.[8]

The skies had been intermittently cloudy all day that September 9, and there had been scattered showers. The launch countdown was proceeding. The meteorologists at the Cape reported that a large storm was rolling in. If we did not launch soon, we would have to scrub for the day. We obviously could not launch during a storm, but we also could not launch into various types of cloud cover. The clouds were intermittently rolling over the Cape as the storm approached.[9]

Jim and the launch team had noted that there were intermittent breaks in the cloud cover. "Can you tell me when we will have a break?" Jim asked the meteorologist.

"Yes, we can give you that time," came the answer.

"Then we will stop the countdown clock at the next possible hold and start it back up to synchronize the launch to the break in the clouds," replied Jim.

The launch clock was stopped and held. The clouds continued to float overhead. The storm was on the way. The cloud break was coming. The countdown clock was restarted. The sky opened up, the clock went to zero, and Viking 2 left the launch pad at 2:39 p.m. EDT September 9, 1975. Jim had shot Viking through a hole in the clouds on the very last day of the nominal launch period.[9]

Less than five minutes later, a very mean storm with high winds, rain, and lightning rolled onto the Cape and caused over two million dollars' worth of destruction, but Viking 2 was no longer there.[10]

Viking 2 was on its way to Mars!

Viking 2 on the TitanCentaur
Ready to Launch
Credit: NASA L 07180

56-A Chance Encounter

During the launch operations at the Cape, I would fly down occasionally to meet with Jim on various matters. One of these trips occurred just a few days before Viking was to be launched. On this trip to the Cape, I happened to sit beside a gentleman named Reginald Turnill. It turned out that Mr. Turnill was the BBC's aviation and space correspondent. I have always equated him as the British equivalent of our Walter Cronkite, at least for space and aviation reporting. We had a great discussion on the way down and made plans to meet once we were at the Cape. This meeting did not happen because of our hurried schedules. A short time later, I received a package in the mail containing two of Mr. Turnill's books, *The Observer's Book of Manned Spaceflight* and *The Observer's Book of Unmanned Spaceflight*. Each book was nicely autographed and contained a short note.

These are still on my desk today.

57-Where Do We Put It?

Viking 1 was now approaching Mars with an arrival on June 19, 1976, after a 304 day journey.

Now the Landing Site Selection (LSS) Team would go into action. Gentry Lee, who was the Science Analysis and Mission Planning Director, was responsible for insuring the science requirements were properly reflected into the mission design. The major science requirements at this point were those associated with finding appropriate landing sites for Viking 1 and 2. As the Landing Site Selection Team decided to look at different sites on Mars, Gentry had to oversee the planning arm of the flight team, which would develop the orbit changes and plan the reconnaissance to get the information about each possible Mars site. Hal Masursky from the US Geological Survey was the Landing Site Selection Team leader, Norm Crabill was the Executive Secretary, Doyle Vogt from The Martin Company, and I were staff to the team.

We had to look at all of our preselected sites which were at highly varying latitudes on Mars and we had to do a lot of the work with Viking Orbiter 1. We had to use Viking Orbiter 1 to look at the A sites, designated for Viking Lander 1 and, if neither the prime or the backup sites were acceptable, we had to find an acceptable site.

We also had to use Viking Orbiter 1 to look at both the B sites (approximately 44 degrees North) and the C sites (approximately 5.5 degrees South) designated for Viking Lander 2. And we had to make a decision regarding the landing latitude of Viking Lander 2 twelve days prior to the arrival of Viking 2. We had to do this because we had to choose the option of going to the B or C sites (different latitudes) for Lander 2 in time to make the last mid course correction, which was dependent on the landing latitude. This last midcourse correction had to be loaded into the spacecraft twelve days before Viking 2 arrived at Mars. Now another race against time began.

Lander-1	A1 19.5° N 34° E	Chryse Planitia
	A2 20° N 108° W	Tritonis Lacus
Lander-2	B1 44° N 10° E	Cydonia
	B2 44° N 110° E	Alba
Alternate Sites	C1 5.5° S 44.5° E	Capri
	C2 5.5° S 3° E	Meridiani

Viking Preselected Prime And Backup Landing Sites

The main instruments for conducting site selection were the two charged couple device cameras on the orbiter that would provide photographs of the Mars surface from which mosaics could be made. In addition to the cameras, the orbiter carried an infrared thermal mapper (IRTM) so we could measure the temperature of individual areas on the Mars surface at different times of day to understand their thermal inertia. If the surface heated and cooled quickly, it could indicate a sandy surface. If it took a long time to heat and cool, it could indicate a rocky surface. The orbiter also carried a water vapor detector to measure the water content in the atmosphere, which could indicate the presence of water near the surface.

We felt that Earth based radar could also help in the selection of the site. As Earth and Mars rotate and move around the sun, there will be times when a radio beam from Earth can be bounced off a small portion of Mars that is directly facing Earth at that moment. This point that faces Earth is called the sub Earth point. Depending on the Earth Mars geometry, this sub Earth point will trace a path over a small portion of the Mars surface. The prevailing logic was that the pattern of the returned signal from the sub Earth point could indicate if that particular surface area is rough, and the strength of the signal would indicate softness of the surface (in which case the radio signal would be partially absorbed by the planet). We were planning to use the 1000 foot (350 meter) antenna at Arecibo, Puerto Rico, which could bounce a signal off Mars that would pass over our landing site on July 4 and 5. This information was to become very useful and provocative.

Also, it must be realized that the instruments we had could not see the surface features at the scale we needed to say a site was "safe." A rock sticking out of the soil nine inches high could penetrate the bottom of the lander. A slope greater than 14 degrees could cause the lander to tip over. And we were viewing the surface with 100 meter resolution. So we could not see anything smaller than a football field. We could measure slopes on a large scale, but we could not see the small, local slopes. This meant that the geologists had to look at the big picture, infer the geologic causes that formed those large features, and then infer what the site would look like at the smaller scale, based on results of similar geologic processes on the Earth and the Moon. Tricky. It was called geologic extrapolation. It didn't work, but it was the best we had.

Now add the possible landing inaccuracies. Because of the lack of complete knowledge of the Mars atmosphere, gravitational field, our map of Mars, and simply the errors in executing the descent and landing maneuvers, our actual landing point could be anywhere within an ellipse 60 miles wide by 130 miles long centered on our planned landing spot. We were more likely to land near the center than the edges of the ellipse but we still needed to find a site that would be hospitable over that entire area. So, the challenge for the Site Selection Team was to find an area on Mars (slightly smaller in area than the state of New Jersey) on which that ellipse would fit with some degree of assuredness that the surface would not be too rough, without too large rocks, and without too steep slopes.

We began looking at our primary planned landing site (known as the A1 site) and realized it was not what we had seen from the Mariner 9 data. We were now looking at Mars with 100 meter resolution instead of the 1000 meter resolution that we had from Mariner 9 photography in 1971.

We could now see the football field sized features instead of features larger than stacks of battleships.

Mars looked totally different with this new, higher resolution. That surface looked rough, even at the 100 meter scale. Also, remember the Mariner photography had been taken looking through the remains of the dust storm, and evidently there was enough dust still in the air to make the terrain appear smoother than it really was.[1]

We began making mosaics with the pictures we had taken. Before long we had 200 to 300 people working at JPL and at the Astrogeology Center in Flagstaff making mosaics, counting craters, and trying to determine slopes. The Flagstaff plane was now making daily flights from Flagstaff to Pasadena to bring the mosaics and analyses. We were also using Amtrak to deliver products between flights. And we were just beginning to realize how much we had to learn about the surface of Mars. Witness the conversation between Hal Masursky and Mike Carr of the USGS, as they looked over some of the early pictures of the A1 and A2 sites.

LSS (Landing Site Staff) Meeting number 6[2]

June 23, 1976, 4 days after insertion into Mars orbit

Mike Carr (USGS): "From the first looks, the A1 site is rough with chaotic terrain, crater fields, and ridges."

Hal Masursky (USGS): "I am not sure if the roughness is peculiar to the site. Frames taken of the A2 site show similar craters and roughness.

The Leaders of the Pack
Shown Left to Right: Hal Mazursky, Gentry Lee, Tom Young and Jim
Martin, listening to the debates over the site.
I am partially seen behind and slightly
to the right of Gentry's head.
Courtesy: Hans-Peter Biemann, Author, *The Vikings of '76*

No matter how you looked at the A1 site, it looked rough. We had to find an area where we could overlay our landing ellipse and feel that we could put our lander safely on the surface within that area.

The problem with the A1 site turned out to be a riverbed that had been cut up and contained many islands, channels, and even craters. We had originally thought that the site was at the mouth of the riverbed, but instead it was in the bed itself. It had undergone a large amount of scouring and appeared to be very rough. Finally, Dr. Bob Hargraves from Princeton (the leader of the Lander Magnetic Properties Team) said, "If you do not like this excavated riverbed area, look for the area where the debris has been deposited."[3] It was such a simple idea, whose time was right.

On June 25, six days after the Viking 1 arrival at Mars, Jim Martin decided to abandon the A1 site. So the question was, do we move westward toward an area that looked like it might contain more hospitable terrain, or do we go to the A2 site that was much farther away, and would take more time? Additional pictures of the A2 site, taken a few days later, looked equally as bad or worse than the A1 site, so the decision was made to look farther west.

The tension stepped up another notch. We had one more chance to find a site before Viking 2 arrived. The last day to land Viking 1 was July 24. From July 26 through August 8, we would have to place Viking 1 in a safe mode and concentrate on getting Viking 2 into orbit about Mars. If we had any problems with this new plan, we might have to land Viking 1 after conjunction when the Sun would go between Earth and Mars, and all communications would be lost for about a month. If this occurred, we could not land at the C sites because it would be too hot. So the constraints mounted. (If you think this is confusing to read, try living it.)

Hal and the Mosaics
Hal Mazursky, going over one of the many mosaics that were
developed to find the best sites for the landers
Courtesy Hans-Peter Biemann, Author, *The Vikings of '76*

With the decision not to land at A1 or A2, we realized we
would not land on July 4 as originally planned and as we had
communicated to the press. At the frequent press
conferences, the reporters were always asking about the new
landing date. The questioning became intense and a little bit
aggressive at one press conference when a journalist asked
Tom Young about the present planned landing date. Jim
Martin stepped up and said, "I'll take that question. We will
land when I believe we have found a safe landing site and not
one minute before." That slowed down the questioning on
thatsubject.

With rejection of the A1 site, we needed a new mapping
sequence to examine the westward possibilities. Hal
Masursky and I were dispatched to develop a new sequence
of photographs that would help us understand that portion of
the Mars surface. We developed the sequence and fed this to
the Mission Planning Group, which would develop the high
level mission sequence of events to accomplish the job.

Later, as I was reviewing this sequence of events, I realized that we had the proper photographic sequence, but we had not changed the readout sequence of the photographs. We would be transmitting unimportant pictures back to Earth instead of the pictures from the new sequence that we needed now.

This would eat up the precious time that we needed to make the mosaics of the new area. I revised the high level readout sequence, and since the preliminary command conference had already been held in which the high level mission sequence of events had been handed over to the Orbiter Performance Analysis Group (OPAG) for command generation, I gave the readout sequence directly to the OPAG. Feeling good about what I had done, I went back to the motel for a short rest.

When I came back into the SFOF, I was greeted by, "Boy, you certainly stirred up the pot." When the final command conference was held and the OPAG presented the set of commands to perform the mission sequence of events that had been given earlier, the commands and mission sequence of events did not match, because I had changed the photographic readout sequence. I had done what was necessary and correct, but I had not informed the mission planning group. I had ignored the very protocol that I had spent much time and effort developing.

Gentry Lee put it very succinctly, "Do we give him a hero button, or do we throw him out of the SFOF?"

Well, they did not throw me out, and I did not get a hero button. I saw Jim shortly after the final command conference, and I really did not know how this conversation was going to go.

Jim looked at me and said," You certainly caused a stir in the command conference. Gentry was very vocal about it."

"Yeah, I heard about that."

"John, you did the right thing. You just did it in the wrong way." He thought a moment and said, "Thanks. But do not do it again."

"You are welcome, and I won't."

As we began to look at the new pictures, we saw that the area to the northwest of the A1 site (which we dubbed A1 NW looked more hospitable. In fact, it looked pretty good. Now we needed the radar data from Arecibo to attempt to confirm the surface smoothness or roughness.

Len Tyler, who was reducing the Arecibo data at Stanford, gave us the results in the morning of July 7. And the results were startling. As the radar tracked across the Mars surface, the returned signal strength was normal. However, as the radar track neared the A1 NW site, the returned signal strength dropped precipitously. After the radar track moved away from the site, the signal strength again returned to normal. This was taken to mean that the site was possibly very rough or soft. Also, the abruptness of the signal drop and recovery could not be understood. After much discussion, Len Tyler made the statement, "Jim, I cannot tell you exactly what this signal trace means, but if you had a huge drill and drilled a hole in Mars over the A1 NW site straight through to the other side, you would get the same reading."

I watched Jim's body language, and I knew he had made his decision. We would abandon the A1 NW site, which is what we did. The decision was made that we would change the orbit and look further west. We finally saw an area that looked much more hospitable from the orbiter camera pictures.

So on July 12, twenty-two days after Viking 1 went into orbit about Mars; after twenty-two meetings of the site selection team; after approximately seven orbit trims to look at different sites; after many twenty hour days for the whole site selection team; and after innumerable orbit redesigns and new sequences and equally long hours by the operations team the final decision was made. We had a site in which the initial pictures looked good. This was to be our site for Viking Lander 1. And the site chosen was at 22.4° N and 47.5° W in Chryse Planitia, approximately 500 miles from the originally selected A1 site.

We now stabilized the orbit over the newly selected site, and turned our attention to the lander.

Mosaic of Landing Site for Viking 1
In Chryse Planitia with Landing Elipse Superimposed
Credit: Reference 12, Page 12

The Landing Site Committee
Voting on the landing site
Courtesy Hans-Peter Biemann, Author, *The Viking of '76*

58-Getting Ready For
The Final Test

During the time the new sites were being examined and the orbit was continually being adjusted, the Viking Lander, its computer, and all the subsystems were patiently waiting. Now the site was chosen. Now it was the lander's turn to be in the spotlight, and the immediate and serious work with the lander began its final phase.

All of the design, development, and test; all of the twenty hour days, all of the weekends, all of the emotions, all of the times away from home at some contractor's plant or test site, all of the times being torn between the family needs and the needs of the lander, and all of the sacrifices, willfully made over the last eight years to be part of this enormous endeavor, would now be tested. This amazing vehicle that we would bring to life for the final test would be our scorecard and our grade. And it would tell us if all of our efforts were up to the task and our sacrifices were to be rewarded. The sequence now began.

3:00 p.m. July 18 , 1976, Viking 1

Lander commands for separation are in the spacecraft computer

All updates are in the command load

Viking Lander to be turned on at 7:30p.m.

Separation minus 30 hours

Based on ground commands, the orbiter performs checks on both A & B Viking lander computers. Both A & B computers check out properly. The orbiter commands the lander Computer A to initiate pre-separation checkout. The lander is transferred to internal power. Computer A takes over command. The only power coming from the orbiter now is directed to the Reaction Control System (RCS) heaters and the terminal engine heaters in order to save the lander's battery power. Now, through the computer sequencing, the lander checkout begins. The gyros are turned on and begin to spin and warm up. They will provide the lander with the knowledge of its orientation in space. The Terminal Descent and Landing Radar (TDLR) is turned on and checked out. It will provide the lander with the information of its horizontal velocity with respect to the Martian surface. The accelerometers that are aligned along the three axes of the lander are tested.

The computer continues to pulse and test the remaining instruments and subsystems that will be needed during descent and entry; the radar altimeter that will give the lander knowledge of its altitude during the major portion of its descent; the relay communication electronics that will transmit information back to the orbiter as the lander makes its journey to the Mars surface; the retarding potential analyzer and the upper atmosphere mass spectrometer that together will measure the composition, structure, and constituents of the Mars atmosphere; and the stagnation and ambient pressure instruments that will provide necessary aerodynamic data.

The instruments that will operate on landing (biology, meteorology, the spectrometers, and cameras) are now checked.

Separation minus 9 1/2 hours

All data has been analyzed and verified by the flight team. The orbiter issues a command to the lander to open the Viking Lander Command Gate so that any separation initialization changes can be made by uplink commands.

Separation minus 3 1/2 Hours

The orbiter goes off the celestial orientation in which it was locked on the Sun and the Star Canopus and goes to inertial reference based on its gyros. It will stay in this configuration until the lander has separated, traveled to the Mars surface, and landed. This allows the lander orbiter relay link to be maintained during entry.

Separation minus 2 hours 31 minutes

The separation initialization sequence begins. The gyros are warmed up and spin initiated and continually monitored. All subsystems are again checked and the data verified by the flight team. This is the last time the flight team gets to look at the lander subsystems prior to separation. After this final verification, the Mission Director, Tom Young, may authorize the separate/go command. This command must be received by the lander at least fifteen minutes before separation, or the lander will go into a hold mode.

It was during the separation initialization phase that we got another look at Jim Martin's unrelenting focus. All during those eight years of Viking development, Jim had kept us all focused on doing those things that would promote the probability of Viking success. He had demanded the truth and only the truth. He had demanded that we concentrate on solving whatever problems were identified and not deviating from developing the best path to a solution. And now we saw that focus again.

Jim was in his room in the middle of the operations floor. The room was about twelve feet on a side, mostly glass, and it contained all of his communications equipment. He was listening to the responses of the various lander subsystem managers, and to Rex Sjostrom, head of the Lander Performance and Command Team. They were reporting on the status of their subsystems. "Gyros spinning up and current normal." "Accelerometers tested and verified." "Radar altimeter on with normal output."

At this time, the red phone on the end of Jim's console rang. This phone was connected directly to the White House with no intervening operators. Jim glanced toward the phone, looking somewhat annoyed. He picked up the phone.

"Mr. Martin?"

"Yes."

"Mr. James Martin?"

"Yes."

"Mr. Martin, President Ford "

"You tell President Ford I can't talk to him right now. We are in the middle of lander checkout. Tell him to call back in five hours, and I will be glad to talk to him then." And he slammed the phone down on the cradle.

He had just hung up on the President.

Now the rest of the story. President Ford called back in exactly five hours. For those of us who had worked for Jim for eight plus years, it was a no brainer.

Of course President Ford would call back in five hours! Jim told him to!

He wouldn't dare do anything else!

Separation minus 14 minutes 30 seconds

The separate/go command has been received. The lander computer A now completely takes over the control of the lander. It immediately slams the gate down on the command link. The lander will accept no commands from the orbiter, the ground, anyone, or anything from this point on.

The lander is now a completely autonomous vehicle. The computer transfers the gyros to internal power from orbiter power. The computer powers on the reaction control system valves so they will be warm when they are needed to orient the lander and perform the deorbit maneuver. The ultra high frequency (UHF) transmitter is powered on for transmission during entry. Both lander pyrotechnic control assemblies are turned on. These will fire many small charges during the descent and landing in order to release portions of the lander no longer needed or to change the lander configuration. The computer makes one final check of all subsystems and tests the subsystems against predetermined values. If the tests were to fail, the computer would abort the separation and go into a safe mode and reopen the command gates. All tests are successful. Computer A performs self-checks and determines that it is functioning properly. It now shuts off the capability to switch to Computer B so that no switchover can occur during the entry and landing. Computer A is now in complete control of the lander throughout the separation, deorbit, entry, and landing phases.

59-The Final Test

During the next few hours the lander will be a living, moving, sensing, decision-making machine, flying into a hostile environment (the Mars atmosphere). The lander will deploy and retract various scientific sensors as it descends in order to better understand the Mars environment and atmosphere. The computer will use various sensors for information, such as: the gyroscopes for orientation in space and the accelerometers to sense the change in velocity of the lander. The computer will determine when and how to orient the lander for the different phases of the entry, descent, and landing. The computer will also decide when to perform various actions, such as turning on and off radars and other instruments. After entry into the Mars atmosphere the computer will use a host of aerodynamic sensors together with the gyros and accelerometers to actually fly the lander, like a pilot flies an aircraft. Using these sensors and the attitude control rockets the computer will hold the tip of the aeroshell (the heat protective nosecone) at a consistent angle to the incoming wind stream to generate the proper amount of lift. During this entire time of entry, descent, and landing, the Lander (controlled by the computer) will be blowing pyros (the pyrotechnic devices, small explosive charges), that will release sensors of various types and will open fuel lines, etc. The computer will also reconfigure the lander by blowing pyros to get rid of parts no longer needed, such as the lander base cover and parachute, and the heat protective aeroshell. The computer will initiate these actions either in response to time signals from the Lander clock as it counts down the seconds or in response to signals from various instruments such as a high altitude mark from a radar. I have captured just a few of these signals and actions in this exciting, exacting and complicated sequence in the table below:

SIGNAL	ACTION
11 seconds after separation signal	Computer commands pyros to fire, cutting three separation bolts. Springs push Lander away from the orbiter.
4 minutes after separation	Computer uses information from the gyroscopes to determine the Lander's orientation in space and commands the reaction control system engines to orient Lander into proper orientation for the deorbit (velocity) maneuver.
7 minutes after separation	Computer initiates deorbit maneuver that will slow the Lander's velocity and place it on a trajectory (path) to the Mars surface.
30 minutes after separation	Having completed the deorbit maneuver, the computer initiates commands for the Lander to go to a preplanned orientation as it coasts toward the Mars atmosphere.
3 hours, 9 minutes after separation	Computer readies the Lander for entry into the Mars atmosphere by turning on the sensors needed for aerodynamic information and to better understand the Mars atmosphere. Computer turns on high speed data system to transmit information back to the orbiter during the descent to the surface.

SIGNAL	ACTION
Lander accelerometers sense deceleration due to entry into Mars atmosphere	Computer now takes control of the Lander as a pilot and begins to fly the Lander through the atmosphere to the surface of Mars. Computer uses gyros, accelerometers, and pressure sensors for information, and commands the Lander to correct angle of attack to generate the proper lift. If the computer flies the Lander with too much lift the Lander will skip out of the atmosphere. If the computer flies the Lander with too little lift, the Lander will come in too steeply, will not be able to slow down, and will crash into the Mars surface.
High altitude radar senses high altitude mark	Computer commands parachute mortar (a large cylinder with a propulsive charge) to fire, thrusting parachute out.
High altitude mark plus 7 seconds	Computer commands pyros to fire, separating the aeroshell (blunt heat protective nose-cone) from the Lander body, as the aeroshell will no longer be needed.
High altitude mark plus 10 seconds	Computer commands terminal descent and landing radar to turn on, providing information to the computer to guide the Lander to the surface.
High altitude mark plus 19 seconds	Computer commands pyros to fire, cutting bolts, and allowing springs to extend Lander legs.

SIGNAL	ACTION
Lander radar signals low altitude mark	Computer commands terminal descent engines to be warmed up to 10 percent throttle settings.
Low altitude mark plus 2 seconds	Computer commands pyros to fire, cutting bolts, which will release the parachute and base cover, as they are no longer needed.
Low altitude mark plus 4 seconds	Computer uses information from terminal descent and landing radar and controls the throttles of the terminal descent engines in order to fly a sweeping trajectory to the Mars surface.
Lander radar senses 5 miles per hour vertical velocity achieved	Computer commands the Lander to maintain the vertical velocity of five miles per hour as it descends to the surface.
Lander leg touches surface and flexes	This signals that the lander is on the surface of Mars. Computer initiates terminal engine shutdown and initiates sequence for first two hours on surface.

Credit: Reference 10: Pages 1-21 to 1-25

At 5:12 a.m. PDT on July 20, 1976, we received confirmation that Viking Lander 1 had come to rest on the surface of Mars, nineteen minutes earlier. This wonderful, life like machine, like something out of Star Wars with all of its popping of pyros, deployment of radars, parachutes, base covers, legs, and antennas was now sitting on Mars. We had successfully landed on that planet for the first time! It was the sensation of being submerged in an unbelievable wave of emotion to see something you have worked on for eight years work so well and perform almost the impossible.

There were cheers, hoots, claps and tears. There were tears on the faces of some who were trying to hide them, while others unabashedly let them show. I was motionless for what seemed like hours. It was almost as if it was too much to believe.

We Were On Mars!

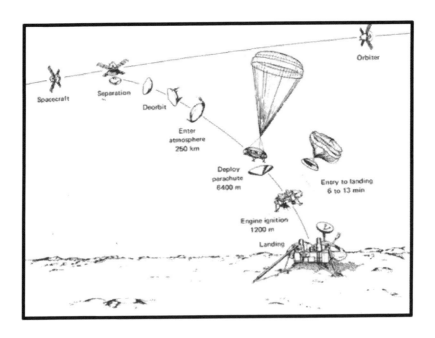

Viking Entry, Descent and Landing Sequence
Reference: 10: page 1-3

60-Viking On The Surface

With Viking on the surface, the computer unfolded and deployed various probes and appendages and put the lander in the configuration shown below in the schematic drawing. One of the first commands was to the high gain antenna to perform a sweeping search so it could locate and transmit directly to Earth.

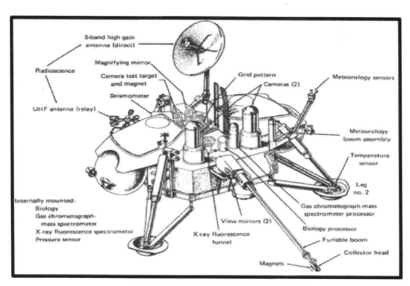

Viking As It Would Be On Mars
Credit: Reference 10: 1-13

Then camera 2 took the first historic picture of Mars from a landed spacecraft. The picture did not show much. It just showed one of the landing legs and some small rocks nearby. It was truly the most beautiful picture I had ever seen.

First Picture From the Surface of Mars.
Credit: NASA LV 00043

The next day after landing we took the first color picture from the Mars surface. This was interesting. The photographs were being received from the tracking stations and constructed in the SFOF photo laboratory. There was evidently a little miscalculation in the setting of the gains (intensity) of the various colors. At first we did not realize that the gains were wrong and were amazed at the fact that the Mars sky appeared blue.

I passed Gerry Soffen in the hall right after this picture came in. He looked at me, grinned, and burst out singing, "Blue skies smiling at me." But the blue sky was not to be. After realizing the error and adjusting the color gains, the sky turned out to be pink, as we have seen in multiple Mars photographs since the initial one. Jim ordered all of these pictures with the blue sky destroyed. Somehow this one got through.

First Mars Picture with Blue Sky
I realize it is difficult to see in black and white,
but it really was blue.
Credit: NASA LV 00045

The Viking lander cameras were interesting in themselves. They were housed in two masts mounted on the lander body. They looked through slits, and a mirror scanned up and down the slits.

They took pictures by scanning down the slit and then rotating the slit a very small amount and scanning down again. It took five minutes, for example, to take the first black and white picture of the Mars surface and the lander footpad shown earlier.

The Viking lander camera team, headed by Tim Mutch, was testing the camera on the western desert (picture next page). You see Tim on the left side of the picture kneeling, and then three more Tims. He had simply gotten up after the camera had scanned his area and walked around to the next spot so he got in the same picture four times, illustrating how slow the process worked.

Viking Lander Camera Picture with Four Tim Mutches
Courtesy of Ed Guinness, Washington University, Department of
Earth and Planetary Sciences,
St. Louis, MO.
Reference: NASA SP 425

61-The Site We Rejected

One further note regarding the landing site for Viking 1. Much later the 1997 Pathfinder project decided to land at the original Viking A1 site that we had rejected. Pictured on the following page is what that site looked like through the eyes of the Pathfinder camera. Mathew Golombek, Mars Pathfinder Project Scientist, reported, "The landing site is among the rockiest parts of Mars. It was the right decision for Viking not to land there."

Score one for the site selection team.

The Rejected Viking A 1 Site
Shown through the eyes of the Mars Pathfinder
Courtesy, Mathew Golombeck
Mars Pathfinder Project Scientist
Credit: NASA/JPL PIA 02406

62-This Thing Ain't Working

Now, Viking Lander 1 was on the surface, and Viking Lander 2 was barreling toward Mars, destined to go into Mars orbit in fifteen days.

The Lander soil sampler arm would not extend.

We still had an unknown factor. Would Lander 2 land around 45 degrees north or 5 degrees south? Remember we had agreed that if Lander 1 was successful, Lander 2 would be free to go to approximately 45 degrees north. If Lander 1 was not successful Lander 2 would have to go to 5 degrees south, which we originally deemed to be the safest area we could find on Mars. Lander 1 success depended primarily on the soil sampler arm working. If it did not work, we could not get any soil samples or perform any of the lander biological experiments, and we could not claim success. We would need to send Lander 2 to 5 degrees south. We had just three days to determine if Lander 1 was deemed a success. This was because we needed to load the final midcourse maneuver sequence for the Viking 2 Spacecraft at least twelve days prior to Mars orbit insertion. The final midcourse maneuver was dependent on what orbit we chose, which was dependent on the latitude of the Viking Lander 2 landing site.

The LSS Team had made the decision to send Lander 2 to the 45 degrees north latitude sites, based on the assumption that Lander 1 was going to be successful. The design of the last mid course maneuver had been made two days before the deadline.

Now, we were at the deadline, and we were faced with a new situation.

Landing Site Selection (LSS) Team Meeting; Number 31: July 23, 1976, Friday, 3:00 p.m. [1]

The team was made aware of the fact that the soil sampler arm would not extend.

The discussion of the landing site latitude for Lander 2 began again with a new possibility that Lander 1 had essentially failed. Everyone had a different opinion.

Hal Masursky; (USGS): We have pictures of the C1 site, and it does not look good. A quick analysis of the C1 site says it looks more hazardous than the A1 site that we abandoned as being too hazardous. And C1 was supposed to be our safe site. We can't see C2 from the present orbit of Viking 1 so we have no data on that possibility. "

Chuck Klein; (Harold Klein, Biology Team): "I do not see that the B latitude is substantially less safe than the C latitude, and the B latitude has substantial science benefits, so for me, the B latitude is the one."

Klaus Biemann; (Biology Team): "If both the B and C latitudes are equally safe, and we have a better chance of finding some evidence of life at the B latitude, then that is the bottom line."

Carl Sagan; (Lander Imaging Team): "I am concerned about no radar at the B latitude.

The worst-case scenario is: Lander 1 sample arm does not work, and Lander 2 crashes. Therefore, we must go to the absolutely safest place possible, and that means a place where we have both radar and photographic data. That says we go to A2 or C1."

Len Tyler; (Leonard Tyler, Radio Science Team): " C1 looks good from the radar. We could get fooled at C2 just as we did at A1."

Gentry Lee; (Science Analysis and Mission Planning Director): "If the arm is not working and we have no place in the B3 area in which we can fit our ellipse, then we should go to C2."

Lou Kingsland; (Deputy Mission Planning Director): "Remember the C2 altitude problem. It looks like C2 could be too high to allow a safe landing."

Jerry Soffen; (Viking Chief Scientist) represented Joshua Lederberg , Biology Team: "The A site, as we predicted, is not biologically interesting. We now have stronger reason to go to the B latitude, where we might have permafrost, which would indicate water."

The hours ticked by. The LSS team debated. Len Clark, who was head of the surface sampler team, gathered team members Don Crouch and Bob Grossart, and together they worked to determine what was wrong with the arm. They finally concluded that the surface sampler arm must be binding against the latch pin that had been holding it in place for the trip to Mars. Len's team was now working with the Martin engineers in Denver who were using an engineering model of the surface sampler. By conducting repeated demonstrations of the original surface sampler arm extensions with the engineering model they were able to show that the short extension command they originally used didn't always guarantee that the latch pin would fall free. At this point Len knew that they could easily fix the problem with a slightly longer extension command.

Jim Martin was heading to the stage for a prescheduled press conference. He knew he was going to get a lot of flack about the surface sampler arm problem. Len beckoned him back as he approached the stage and simply said, "We can solve the problem". As Jim took his seat for the press conference, the first question that was asked, was, "Mr. Martin, we hear that the surface sampler arm is not working and can not deliver soil to the biology instruments. Does this mean that Viking 1 has failed?

Jim, armed with the last minute message, would then calmly say, "These are normal glitches that you get on any space mission. We can solve the problem. The primary mission is not in jeopardy."

Len immediately went back to the operations team with the surface sampler team's finding, and on sol 5, with a new set of lander commands, the surface sampler on Lander 1 moved and the latch pin fell to the Martian surface. The surface sampler was now working again, and Lander 1 began retrieving soil for the experiments. Lander 1 could now be considered a success.

On the last day to determine the landing site latitude for Viking 2 just in time to make the last midcourse maneuver, we were able to make it, and we set the course for landing at approximately 45 degrees north. We had used every day we had in the separation of the two spacecraft arrivals at Mars in order to find a landing site for Viking 1 and for solving the sampler arm problem. Viking 2 went into orbit about Mars on August 7, 1976 after a 333 day interplanetary journey.

I have reflected many times on that period in the mission operations. If we had not opened up the separation in arrival times at Mars for the two spacecraft to 50 days, from the 30 that we originally planned, Lander 2 could very possibly be sitting at 5 degrees south today, instead of the desired site, which actually turned out to be over 47 degrees north. As we designed our mission, we certainly did not know that we would take thirty-two days just to get the first lander on the surface, and then the additional days to get the surface sampler arm freed, leaving us just barely enough time to make the last midcourse maneuver for Viking 2. I do not know what the resulting mission would have looked like if the surface sampler team had not solved the surface sampler arm problem when they did, but it probably would have been a much different mission.

A Heartfelt Congratulations
In the picture, Jim Martin is shaking the hand of Len Clark after his team's discovery of the soil sampler problem.
Also shown next to Len is Tom Young
Mission Operations Director.
Courtesy Hans-Peter Biemann, Author, *The Vikings of '76*

63-Where Do We Park Viking Lander-2?

While the lander team was working madly to get the soil sampler operational and beginning to perform the set of landed experiments, the Site Selection Team was busy trying to find a suitable site for Lander 2 at the 45-degree north latitude. Once more, we were fighting the clock. We wanted to get Lander 2 on Mars rapidly so the lander could operate a minimum of sixty days on the surface prior to conjunction. The Mars Sun conjunction would occur in early November, at which time Mars would go behind the Sun, and the communications to and from both the landers and orbiters would rapidly deteriorate and then drop out. We hoped that we could pick up communications about a month later as Mars moved out from behind the Sun, but the prime mission for the Vikings would be over when we lost the signals. It was imperative that we find a good site for Viking 2 and land as soon as possible.

The more we looked at the prime site for Lander 2, designated B1, the rougher it looked. We abandoned B1 and reexamined B2, which we finally rejected. Again, we began looking for alternate sites. Before long we were considering five different sites ranging from 43 degrees to 48 degrees north, and from three widely different locations approximately 120 degrees apart in longitude.

The flight operations team, which was still trying to recover from the site selection activity for Lander 1, was developing and maintaining mission sequences for all of these different possibilities and was near exhaustion. More pictures were coming in, and the mosaics were being developed. There were some indications that two of the sites might have sand dunes, which raised the question: which had the most sand coverage that would possibly cover large rocks that could, if exposed, kill the lander?

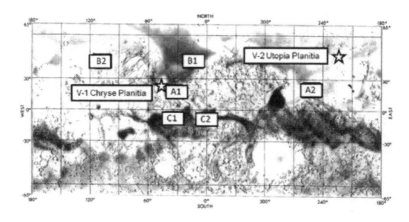

Mars Map Showing Landing Sites
A1 through C2 show planned sites;
V 1 and V 2 and Stars show actual landing sites
Map Courtesy of Viking Flight Operations Team

Landing Site Selection Team Meeting; Number 46: August 21, 1976; Thursday, 4:00 p.m. [1]

Tom Young: "We have run out of time. We must specify the landing site coordinates by tomorrow at 8:00 a.m. This will define the trim maneuver we will make on Wednesday, the 25th. We have two options at this point. We can continue to plan for the site we have named B3 East, or we move back to B2 West. We must determine which area we feel is safest."

Jim Martin: "I do not plan to land if it is not safe. However, the current plan is to land September 3rd."

Gentry Lee: "Remember that we canceled the B2 options. It would be a six to seven day impact to pick up that option now."

Conway Snyder; (Orbiter Chief Scientist): "Have we changed the approach we said we would take last night? Then we said we would choose from the best of B3 East or B3 West. Now it looks like we are considering B3 East or B3 West or punting back to the B 2 area."

The answer was yes; the B2 option was still open.

Hal Masursky: "In developing the candidate ellipses (possible landing areas) for both the B2 and the B3 sites, we had to look at areas not acceptable by our previous criteria."

Hal Masursky: " B2 craters are small. B3 craters are large. Our confidence in B3 is less. We can clearly identify big dunes in B2 that can hide the smaller craters that exist there."

It was now beginning to look like both the B2 and B3 areas possibly had sand dunes in some of the areas.

John Guest (USGS): "There seems to be good evidence of dunes at B2. We are not as confident about dunes at B3. However, there was cloud cover at B2, so that may account for the appearance of dunes."

After much discussion and review of the new information, a vote was taken. Twenty were in favor of B2 West and ten for B3 East. These votes gave a sense of the team's views.

In reality, only one vote counted, and Jim Martin had it.

Landing Site Selection Team Meeting; Number 46: August 21, 1976; Executive Session; Thursday 8:00 p.m.[2]

Hal Masursky: "I am in favor of B2. We believe sand dunes are there. In B3, we can't estimate the depth of the sand coverage."

Mike Carr: "We have much greater confidence in the geology model at B3. All factors merge to indicate dunes over bed rock."

Tom Young: "With regard to safety: it appears that B3 has a cover of sand over the surface. The experts are not as certain of B2. With regard to science there is only a small distinction. Regarding operations: implementation is much more straight forward at B3."

Jim Martin: "To go to B2 and get new data could take up to two weeks. If we went to B2 and were wrong and could not find a suitable landing spot, it is very likely we could not complete our mission before conjunction."

There was much more discussion. We had a lot of data that implied what the surface might be like. But we had nothing that really told us what the surface would be. We had to extrapolate what the surface might be like and try to maximize our probabilities of landing in a "safe" area. Finally Jim announced, "We are going to B3. It is safe enough. It has good science. It has no radar observations, but the project is willing to take that risk."

So the decision was made. Tom, Jim, and Hal, determined the coordinates of the site for Lander 2 to be 48 degrees north and 226 degrees west.[3]

At first, Hal was concerned about the choice of the site. His comment was, "We are going to a site about which we have no data." We had some early, oblique photography of the site, but not the detailed photography we really wanted. However, Jim was keenly aware that time was running out. Also, the flight operations team that had been maintaining numerous options regarding the different sites was near exhaustion. He also felt that this was the best possibility.

We had seen enough to plan on going there, and the later pictures would hopefully confirm that. If they did not, the search would have to continue elsewhere.

As the orbiter pictures came in and the mosaics were made, it began to look like this would be the site. The orbit was stabilized on the new site in Utopia Planitia on August 27th, and preparation began for the next lander's descent to the Mars surface seven days later on September 3rd.

64-Let's Do It Again

The operations team now went through the same process as with Viking 1 reading out and verifying all the information in the lander in order to ensure the proper command loads were in place and ready for execution, plus verifying all the final tests of the lander's subsystems. All went well, until lander separation.

Then, shortly after lander separation, the orbiter gyros shut down, and the orbiter began to drift off its proper orientation in space. This caused the orbiter high gain antenna to lose lock with Earth, so no data could be fed back to Earth from the lander through the high gain orbiter antenna. The only lander data we had now, as Lander 2 descended to Mars, was that coming from the lander low gain antenna back directly to Earth. Instead of the full stream of data we had from Lander 1 as it descended, we now had a thin stream of information that barely indicated to us that Lander 2 was proceeding properly. All eyes were glued to the monitors watching that thin stream of data. It was like holding on to Lander 2 with a thin piece of thread.

Now the time came for the lander to reach the Mars surface.

The time came and went. The signal (a bit rate change in the data stream) that the lander had landed safely did not arrive.

All hearts stopped. All breathing stopped. It was absolutely and eerily silent in the SFOF. The only noises came from the occasional clicks of a data recorder somewhere in the area.

We waited. It seemed like days. All eyes were glued to the data monitors, waiting for the indicated signal. No one moved. It was like the whole world stopped. Finally, over a minute later than expected, on September 3, 1976 at 3:58:20 p.m. PDT, we saw the bit rate change and we knew Lander 2 had successfully settled on the Mars surface, nineteen minutes earlier.[1]

There were the same hoots, cheers, hollers, and tears that accompanied the first landing.

Congratulations!
Noel Hinners (NASA Associate Administrator for Space Science) congratulating Jim Martin, left, on the second successful landing
Courtesy: Hans-Peter Biemann, Author, *The Vikings of '76*.

We had now done it twice.

We had two landers on the surface of Mars, and after some commanding of the orbiter, we were able to get the gyros working again, restore the orientation in space, and resume the high speed data link from the lander through the orbiter back to Earth. All was well with the world!

No Hang-up This Time!
John Naugle and Jim Martin receiving congratulations from
President Ford after second landing.
This time Jim did not hang-up on President Ford.
Courtesy: Hans-Peter Biemann, Author, *The Vikings of '76*.

One interesting factor remained when we looked at the first picture of the Viking 2 landing site. Some of the rocks in the picture are 4 to 5 feet across. So much for the sand dune theory. Hal Masursky was presented with a picture of the site, and was reminded that he had thought it was going to contain sand dunes. Jim, after presenting him with the picture, said laughingly, "The next time I want a landing site, I will ask for one with rocks."[2]

We kept relearning how little we could determine of the real Mars surface from the 100 meter photography.

The view of Utopia Plantia
From the camera of Lander 2
Credit: NASA P 17690

A Moment of Relaxation.
After the second landing
Looking at the Lander pictures.
Credit: NASA JPL P 17924B

In the photograph above, I am in the center with a white shirt and tie. On my right in the black jacket is M. J. (Al) Alazard, Director, Mission Control. To my left and in the foreground is Al Shallenmuller and John Adamoli, both deputy directors, Mission Control.

65-How Viking–1 Opened The National Air And Space Museum

While those of us on the West Coast were trying to get the two Viking landers on the surface of Mars, a small number of Vikings took a few days out and brushed shoulders with the President and associated friends.

Joe Gowdey related the story: On July 2, 1976 many dignitaries were gathered on the patio in front of the main doors leading into the new National Air and Space Museum.

The Air Force Band had opened the ceremony, and the Thunderbirds had executed a flyover. The colors had been presented, and the National Anthem had been performed. William Creighton, Bishop of Washington, had given the invocation, and S. Dillon Ripley, Secretary of the Smithsonian Institution, had welcomed the guests.

Other notables on the patio that morning were Nelson Rockefeller, Vice President; Carl Albert, Speaker of the House; and Mike Collins, Command Module Pilot on Apollo 12, who was now the Director of the Air and Space Museum.

The honorable Warren E. Burger, Chief Justice of the Supreme Court, now introduced President Ford, who would give the main address opening the museum. Viking 1 had gone into orbit about twelve days earlier on June 19, 1976. It had been decided long before Viking went into orbit that the Viking Orbiter would actually send the radio signal that would open the museum. This signal would cause a model of the Viking lander soil sampler arm to cut the opening ribbon.

The plan was that the orbiter would send signals every minute to the tracking station at Goldstone, California. These signals would be transmitted through the operations facility at JPL to the Vikings behind the front doors at the Air and Space Museum. At the proper time they would let the signal through to a model of the Viking lander surface sampler arm that was mounted on the front patio of the museum. The arm, which Vernon Gillespie modified for this purpose, had been outfitted with a hot wire that would close on the ribbon and burn it in two. Leo Daspit, who headed the team, had also gotten a traffic light from the city of Hampton, VA to use in the countdown to the ribbon cutting.

The orbiter was now sending a special pulse every minute on the minute. Joe Gowdey, who had handled the electronics, let the first pulse through, and the traffic light turned on and glowed red. At the time President Ford strode to the podium, he had exactly ten minutes to complete his remarks. The pulses continued to come each minute. The President continued to speak. Finally the nine minute pulse came and the traffic light turned yellow, meaning the President had one minute to finish his speech. President Ford did his part beautifully. The final pulse came down. The light turned green. Joe Gowdey had been watching the ceremony through the front doors of the museum. He let the pulse through to the surface sampler arm. The wire heated. The arm moved back and down. The hot wire caught the ribbon and burned it in two, right on schedule, opening the Washington Air and Space Museum for all to enjoy.

A photograph of the program for the opening of the National Air and Space Museum is on the next page. Also a portion of the ribbon is shown in the upper right hand corner.

The Inscription Reads:
To Joe C Gowdey
With great appreciation for getting
Viking to open our museum.
Mike Collins
Courtesy: Joe Gowdey

66-Lander Surface Experiments

Once on the surface, the landers had a host of experiments to perform. The most widely published experiments involved the pictures taken by the two Viking cameras giving us the first up close pictures of the Mars surface.

There was also the physical properties experiment that measured features such as the surface strength and soil penetration resistance.[1]

The magnetic properties experiment measured the magnetism of the different types of Martian soil and in this way determined its level of oxidation.[2]

The meteorology experiment kept giving weather reports from Mars with temperature, pressure, wind speed, and direction. Because of the length of time the landers remained active, we were able to record atmospheric conditions over more than one complete Mars year, leading to the discovery of diurnal cycles or "seasons" on Mars.[3]

We also found nitrogen in the atmosphere, which was exciting because it is a necessary component of protein molecules that form living things.[4]

The seismology experiment used Mars quakes to provide information such as the thickness of the Mars crust as well as other features.[5]

The radio science experiment helped determine the actual

spin axis of Mars and the actual Mars ephemeris. (Ephemeris is the position and velocity of Mars as it travels around the sun.) Later during conjunction, as Mars went behind the Sun, the radio science team was also able to perform an experiment regarding Einstein's theory of relativity.[6]

The inorganic chemistry experiment was able to identify and quantify many constituents of the Martian soil.[7]

The orbiters continued to photograph the Mars surface and map the surface temperature and water vapor contained in the atmosphere. We saw clouds. The Mars atmosphere contains only a small amount of water, but it is enough to condense under certain conditions and form clouds.[8]

From the lander entry and surface experiments, we gained much knowledge of the chemical composition of the Mars atmosphere.[9] The largest portion of the atmosphere turns out to be carbon dioxide (approximately 95 percent).[10] We also learned that the permanent polar caps are water ice. The polar caps that grow and recede are carbon dioxide.[11]

67-Is There Life On Mars?

We were truly beginning to rewrite the book on Mars. We were learning about the geology, the constituents of the soil, the physical and magnetic properties, the Mars seismic activity, the atmosphere, and the weather. However, the most pressing question remained. What about life on Mars?

On sol 8, the soil sampler made its first dig and delivered soil to the biology instruments and the organic compound analysis instrument, the gas chromatograph mass spectrometer (GCMS). Now the experiments began, and so did the discussions that were to go on for years. There were many intermediate results from all instruments, some of which caused very animated discussions.

Let me summarize the experiments performed, the arguments, and the conclusions that were reached at that time regarding the presence of life and the question that still lingers today.

We had three experiments that were designed to detect life in different ways. These were the Gas Exchange Experiment, the Pyrolytic Release Experiment, and the Labeled Release Experiment. As mentioned, we also had a gas chromatograph mass spectrometer to detect organic materials essential to life as we know it. This is what happened:

We now began the Gas Exchange Experiment (Principal Investigator Dr. Vance Oyama). This experiment is based on the idea that all living things, plants and animals, give off gasses. "Plants give off oxygen, animals give off carbon dioxide, and both exhale water."[1] Therefore, we would place a sample of the soil in the incubator and provide it first with water vapor only, then liquid water, and then a complex nutrient aqueous solution. Then we would see if the composition of the gas above the sample changed, which would indicate some type of metabolization was present.

A soil sample was delivered to the gas exchange experiment incubator, and the sample was exposed to humidity alone, then water, and then nutrients. There were initial pulses of oxygen and carbon dioxide, which caused a lot of excitement and heated discussion. But after more analysis, it was agreed that the results were due to reactions with the Martian soil and not the result of any metabolism.[2] So the conclusion from this experiment was: no life detected.[3]

The Pyrolytic Release Experiment (Principal Investigator: Dr. Norman Horowitz) was designed to look for photosynthetic processes, in which the plant type organisms (if there were any) would take carbon monoxide (CO) and carbon dioxide (CO_2) from the air and convert it into the plant material. In this experiment, a soil sample was placed in the test chamber along with the normal Martian atmosphere and a xenon lamp to simulate the sun. Radioactive CO_2 and CO were released into the test chamber, the light turned on, and the sample incubated for 5 days.

The sample was then heated to a temperature to pyrolyze (char) any organic material. The vapor was then swept through a chamber called an organic vapor trap. The radioactive material that had not chemically combined continued through the trap and flowed to the exhaust port. The trap was then heated to a higher temperature to release any organic compounds that had been trapped. The resulting gas was then swept past a radiation counter to see if any radioactivity was present. If so, this would indicate that the CO and CO_2 had been taken up or "fixed" by plants, indicating life had been found in the soil.[4]

Now we had a dilemma. "Seven of the nine pyrolytic tests executed on Mars gave minimally positive results (in that a small amount of radioactivity was detected).[5] After much discussion, it was felt that it was not clear whether the results were due to some microorganisms in the soil or due to a non-biological reaction. Finally, in Dr. Norman Horowitz's words, "The findings of the pyrolytic release experiment had to be interpreted nonbiologically."[6]

The third life detection experiment was the Labeled Release Experiment whose experimenter was Dr. Gilbert Levin with co-experimenter Dr. Patricia Stratt. In this experiment, soil samples were taken from the surface of Mars and placed in a test chamber. A tiny drop of a solution of Miller Urey compounds labeled with radioactive carbon was injected onto the center of the soil samples to supply a gradient from wet to moist soil. The gas above the soil was continually monitored to detect any radioactive gas that might be given off. If radioactivity was detected, it was an indication that some form of life had metabolized the compounds. In that case, an all important control was run in which a duplicate soil sample was heated to a temperature thought to kill microorganisms but not destroy any oxidant chemical that might have produced the result. If the control tested negative, that was the evidence for life. In all four of these experiments, the radioactivity was detected in the gasses, indicating that some biological process had occurred. Five control runs were made, and all indicated that the positive responses were from life.[7]

Viewing this experiment by itself, one could make the strong argument that life had been found.[8]

Next we looked at the results of the Molecular Analysis Experiment. (Principal Investigator Dr. Klaus Biemann) In this experiment, the soil samples were heated to 200 degrees C, then to 500 degrees C, and then fed to a gas chromatograph mass spectrometer (GCMS) that looked for organic compounds. The bottom line, very simply is: none were found.[9]

As this set of data was being gathered during the mission, the debates started. We had one life detecting experiment that showed negative results. We had another experiment that showed weak positive results deemed non-biological. We had one experiment that showed positive results every time, and when the control experiments were run they confirmed the earlier positive results. But we did not find signs of any organic material from the GCMS. And the question kept being asked, how can you have detected life and not see any evidence of organic material? There must be a problem with the Labeled Release Experiment results if no organic materials were found. Several possible scenarios began to be formed as to how the Labeled Release Equipment could possibly show "false positive" results.

The Viking science team mulled over these findings for months and finally came to the conclusion that we had found "no conclusive evidence of life." Gerry Soffen, the Viking Project Scientist, was heard to say, "That's the ball game. No organics on Mars. No life on Mars."[10]

Norman Horowitz (the Experimenter of the Pyrolytic Release Experiment) came to the opinion that there is no life elsewhere in the solar system.[11]

A formal review was held in the fall of 1976 to examine the results of the Labeled Release Experiment.[12] The conclusion that came out of that review and associated discussions was that "no conclusive evidence of life was produced."[13] That, along with many associated discussions, has been the official position of NASA and most of the science community since that time.

Perhaps when reviewing this conclusion, it is useful to examine the mindset of some of the major parties involved. Both Dr. Klein, who was head of the Biology Team and Dr. Horowitz felt that life could only exist within a very limited set of conditions.[14] This belief is based somewhat on the Antarctic soil studies that began in the mid 1960s as experiments were being developed to look for life on Mars. Dr. Horowitz was the first to encourage studies of the dry Antarctic valleys in order to examine those soils for evidence of life. He felt that the conditions there (dry and cold) would be as close as we could get on Earth to the conditions that he felt would exist on Mars. In 1965 Dr. Cameron, who had been working with Dr. Horowitz, went to Antarctica and collected numerous samples from the dry valleys.

When Dr. Horowitz tested these samples, he found no evidence of life. In fact, he and Dr. Cameron, et al, published a paper entitled Sterile Soil From Antarctica: Organic Analysis.[15] Based on these findings, Dr. Horowitz believed that life could not exist on Mars. "His reasoning was based on the presumption that if most Antarctic soils did not have life, then surely Mars would not."[16]

Dr. Wolf Vishniac, who was originally on the Viking Biology Team, held the view that life could exist under very extreme and varied conditions. After reading Dr. Horowitz's paper, he decided to go to Antarctica and also collect samples for testing. In 1971, Dr. Vishniac went to Antarctica's Wright Valley Asgard range of Victoria Land and took samples from the same areas that Dr. Horowitz had previously examined and found to be sterile.

When the new samples were examined with Dr. Vishniac's and Dr. Levin's instruments, they were found to contain living microorganisms in the form of bacteria, fungi, and algae.[17] These results were then challenged by Dr. Horowitz, who claimed "that the microorganisms Vishniac found did not normally live in the soil, but had been dropped there by the winds from the more favorable climates just before Vishniac found them."[18] Given this challenge and a continuing desire to study microorganisms in extreme climates, Dr. Vishniac decided to go back to Antarctica in 1973 to take more samples from the Asgard Mountain Range.

His expedition was going well until the fateful day of December 10, 1973. As he reached over a cliff face to retrieve a sample to be tested in Dr. Levin's Labeled Release Experiment (the same experiment that was to fly on the Viking Spacecraft), he lost his grip on the icy surface and fell to his death 1000 feet below. Ironically, he fell to his death into the valley between Mount Baldr and Mount Thor, named for two of the Viking gods.[19]

68-The NASA Mars Conference

Fast-forward to ten years after the Viking landings to the NASA Mars Conference held in July 1986. Here we heard the two major arguments for and against life on Mars.

Dr. Norman Horowitz, who at that time was Professor Emeritus at the California Institute of Technology and was the Experimenter of the Pyrolytic Release Experiment, summarized his presentation on the Viking Mission.

> *"It is sometimes said that the results of the mission were ambiguous as to the question of Martian life. I believe that this opinion is mistaken. Indeed, it seems to me the results are quite clear. And my reason for reviewing them here is to demonstrate this fact."*[1]

Dr. Horowitz continued by saying,

> *"Viking not only found no life on Mars, it showed why there is no life there. The two sites sampled by the Viking landers were 25 degrees apart in latitude and on opposite sides of the planet. Yet despite this separation they were very similar in their surface chemistry in both organic and inorganic terms."*[2]
>
> *"Viking found that Mars is even dryer than had been previously thought. The equatorial latitudes where temperatures rise well above 0*

> *degrees C and where, for this reason, conditions were thought to be especially favorable before Viking, were found to be desiccated (the driest part of the planet). The dryness alone would suffice to guarantee a lifeless Mars; combined with the planet's radiation flux, Mars becomes almost Moon like in its hostility to life."*

Dr. Gilbert Levin of Biospherics Incorporated was the Experimenter for the Labeled Release Experiment. His presentation provided results of analyses that attempted to explain all of the scenarios that had been developed. He attempted to refute the claimed "false positive" Labeled Release Experiment results. He said:

> *"We have waited ten years for all of the theories, experiments, and results produced by the many scientists investigating our experiment to be reviewed before voicing a committed conclusion of our own. After examining these efforts in great detail, and after years of laboratory work trying to duplicate our Mars data by non-biological means, we find the preponderance of scientific analysis makes it more probable than not that living organisms were detected in the Labeled Release Experiment on Mars. This is not presented as an opinion but as a position dictated by the objective evaluation of all relevant scientific data."[4]*
>
> *"In conclusion, then, we submit that this real possibility for Martian biology should be an important, even dominant, consideration in the future exploration of the planet Mars. This*

> *clearly is not the case at the present time,*
> *according to published NASA plans for*
> *continuing the unmanned exploration of Mars*
> *that neglect the biology issue. The search for*
> *life on Mars, when evidence for its possible*
> *existence offers such important potential,*
> *should have much greater significance for the*
> *planning of such missions.*"[5]

There is still a vigorous debate about the question, "Did Viking find life on Mars?" Did Dr. Gilbert Levin find life with the Labeled Release Experiment? What about the gas chromatograph mass spectrometer finding no evidence of organics?

More information on the question of life on Mars has continued to be revealed. The real analysis and answer will have to come from a new scientific re evaluation of all the data from Viking and from all that has been found since.

And the beat goes on.

69-The Remaining Question Of Life On Mars

Fast forward again to today. We now know many things that we did not know when the Viking experiments were performed. We now know that the gas chromatograph mass spectrometer (GCMS) that was on Viking was not sensitive enough to detect low concentrations of organic materials. The Viking GCMS has been tested on the same Antarctic soil along with the Labeled Release Experiment. The Labeled Release Experiment was able to detect life in the soil, and the GCMS was not able to find any organics.[1] We now know there is water in the Mars soil, the absence of which Dr. Horowitz claimed made life on Mars impossible.[2]

We also know that the Mars Science Laboratory (MSL) has found organics with its mass spectrometer. This result is problematic, though, since the MSL was not as highly cleaned chemically, nor sterilized, as was Viking. The MSL was cleansed "by alcohol swabbing of the exterior of the spacecraft."[3] So the question remains: did the MSL discover organics, or bring them?

However, NASA recently announced that not all of the simple organics that have been detected could have been terrestrial contamination, and that some of the large amount of carbon dioxide given off when the sample was heated could have come from complex organics.[4]

We also have another piece of data from the Mars Science Laboratory (MSL) that may prove to be interesting. Findings from the MSL and other Mars landers indicate that there are reasonable amounts of perchlorates in the Martian soil. The presence of these perchlorates could very likely be responsible for the failure of the gas chromatograph mass spectrometer on Viking to detect organics. This failure was the main argument against the positive results of the Labeled Release Experiment.

So the big question that Viking attempted to answer is still open. It still provides fodder for much discussion among scientists today and possibly for years to come. Possibly the new results from the Mars Science Laboratory may give new emphasis to the discussions.

We will have to wait and see.

70-The Mission Beyond The Mission

Long before the two Vikings were on the surface of Mars, Jim had assigned me as his Deputy for Planning the Viking Extended Mission. This would begin after we reacquired the signals from the Viking orbiters and landers following solar conjunction (when the Sun would come between Earth and Mars causing all communications to cease). We had the basic plans in place, but the contracts had not been executed, since we did not know for sure up until now that there would be an extended mission.

After the landings, Jim called me in and said, "We need to get all of the contracts for the extended mission negotiated and in place."

"Fine," I said. "How long do I have?"

"One month," was the answer. Typical Jim Martin schedule. Well, we had over 20 contracts between us and JPL, the Deep Space Net, Martin Marietta, and then the smaller ones with all of the science teams.

"I need Bob Murphy," I said. He was our cost control guy, and an excellent one, I might add. He had been with us throughout the entire Viking development process. He knew where the money was and how much there was.

I knew I also needed someone familiar with the Deep Space

Net (DSN) that includes the tracking stations all over the world and much of the operations center at JPL.

"I need Marshall Johnson; I can handle everyone else, but the DSN can pull the wool over my eyes, and I will not know it."

"No problem with Bob. You can have Marshall if he will agree to help."

I wondered why Jim answered in that manner. I had not seen Marshall in a while due to the different jobs we were doing, so I was not prepared for our meeting. Marshall was one of those bigger than life guys. He was tall and well built, with a square cut chin, one of those guys that Hollywood would love to have. In fact, Marshall had acted in some films in his younger days and had even done some directing before he went into the space biz. He also knew the DSN like the palm of his hand. He had helped set up the DSN in the beginning, and he had been an integral part of both the Lunar Orbiter and Viking projects. He had also helped set up the flight team and had made almost all of the arrangements with the DSN. I went directly to Marshall's office.

"Marshall, I need your help in negotiating with the DSN for the cost of the extended mission. I need to have all the contracts in place in one month."

Marshall looked up at me, and I have never seen him look so tired. He had always been a tiger, ready to pounce on anyone who dared to say something he thought was wrong. But once you gained his respect, he was your friend and partner. He had literally come across the table at me the first time we met, but we quickly became good friends, with the accompanying mutual respect. Now I was asking him for his help. He pushed his chair back and looked at me for a long time. "I wish you had not been the one who asked. If it were anyone else, I would have told him or her to go to hell. I am tired, John. I am tired of being a son of a bitch. And doing what you ask, I will have to be a son of a bitch one more time."

He stopped and thought for a few minutes. "Damn it! Okay! But let's get it over with!"

So off the three of us went (Bob, Marshall, and me), to get the extended mission under contract. We met Jim's schedule and got all the participants under contract and ready for the extended mission.

71-A Bit Of Nostalgia

The completion of the extended Mission planning was my last major effort for Viking on the West Coast. I would be Jim's Deputy for East Coast Operations and coordinate any efforts required between the two coasts. I was now on the plane headed home. I would only be back in the Flight Operations Center for several more short visits for Viking. We were approaching the Viking extended mission time period. Some of the friends I had made would now be continuing to perform the flight operations for Viking, but many would go on to other challenges. I was going to miss them. I was going to miss the technical debates, the trust and respect that had developed between myself and the other members of the team, the discussions during dinner about Mars, and even the arguments we had, but most of all the feeling of working together to achieve something that none of us could have achieved if all the others were not there. In a way I was leaving a way of life I had known for over thirteen years, including Lunar Orbiter and Viking. There had been many challenges, disappointments, accomplishments, and friendships. I was now headed to an unknown future, but making that trip with a major sense of accomplishment.

Many times a project will produce a pin that states *2 for 2* or something similar, meaning that the project had two launches and two successes. Those of us who had worked on Viking and Lunar Orbiter could now wear a badge that said *9 for 9*, for five Lunar Orbiters, two Viking Landers and to Viking Orbiters. I do not know of any other set of people who worked on space missions who can wear a badge even close to that.

Not many years ago, a twelve year old student out in the Midwest was in a class in which the teacher was talking about space exploration and Mars. The student raised his hand and said, "My grandfather's name is on Mars." There were chuckles in the room. The teacher was taken back but tried not to show it and embarrass the young fellow. "It's true," he said. "My grandfather signed a card that went on the Viking landers." More chuckles. But now the teacher had something to work with. She started calling various individuals and finally, through a series of steps, got the name of Gus Guastaferro, Jim's Business Manager. Gus asked a few questions about where the fellow's grandfather worked during those days, what type of work he did, what was his name, etc. Gus finally came to the conclusion that it was a very high probability that his name was on Mars.

The reason that the grandfather's name was on Mars can, in a large way, be attributed to Carl Sagan. Carl had proposed that all of the people who worked on Viking, along with their spouses, sign a large card, and that an image of the card be miniaturized and placed on each of the two Viking landers. Therefore, my name, my wife's name, the name of the grandfather of that little fellow in the Midwest, and the names of all of us, are sitting on the Mars surface today with the two Viking Landers. This was also a point of pride for my grandsons, who, like the young fellow in the Midwest, announced in their classrooms that their grandparent's names were on Mars today.

Those names sitting on Mars, the memory of the tough challenges and solutions in both Lunar Orbiter and Viking, the friendships that were made, and the fact that many of us can wear a 9 for 9 pin have continued to be a source of pride for me for many years.

And one more note: We were evidently able to prove ourselves through the success of the Lunar Orbiter Project and the early days of Viking, for one of our investigators on the Molecular Analysis Science Team was none other than the highly regarded geneticist, exobiologist, and Nobel Laureate, Dr. Harold Urey.

You will remember that his letter to the NASA Administrator regarding NASA Langley managing the Lunar Orbiter Project provided the name for this book.

72-Viking Epilogue

The first Viking spacecraft arrived at Mars on June 19, 1976 after a 304 day journey of over approximately 370 million miles. Viking 2 arrived after a similar journey of 333 days. The prime Viking mission ended at the Sun Mars conjunction on November 15, 1976. However, the landers and the orbiters began communicating with Earth shortly after the end of the conjunction and began the extended mission. The two landers lasted well past their designed lifetime of 90 days, with Lander 1 continuing to operate over five and one half years and Lander 2 operating for approximately three and one half years. Orbiter 1 and Orbiter 2 operated in Mars orbit for approximately four years and two years, respectively. All spacecraft returned large amounts of highly useful data for their entire lifetimes.

Life After Viking

*Some of the unsung heroes
highlighted in this book*

Clifford Nelson

After Lunar Orbiter, Cliff was appointed Director for Space at Langley and served in this position until his retirement in 1972. After that he had time to enjoy his favorite sport, sailing. He and his wife, Helen, moved to Florida and lived there until his death in 2004 at age 89. We will all remember his "Atta boy's." Cliff did a lot of "management by walking around." He would come upon a small group of us talking and working out some problem and listen for a few minutes to see that we were making progress. Then he would lightly pat one of us on the shoulder and say "attaboy" and walk off. Somehow this motivation never grew old, no matter how many times he did it. We all miss his "attaboys."

Jim Martin

Jim resigned from NASA in late 1976 near the end of the Viking primary mission and accepted the position of Vice President of Advanced Programs for Martin Marietta. He worked in that position until 1985, when he retired. After retiring, Jim continued to consult with NASA, and in 2000 he was called on to review and restructure the NASA Mars Program after the failures of the Mars Climate Orbiter and Mars Polar Lander. Jim continued to work with NASA until his health began to fail, and he finally succumbed to cancer in 2008.

Tom Young

After leaving Viking, Tom became the Director of the Planetary Program in the Office of Space Science at NASA Headquarters. He then traveled across country to become the Deputy Director of the Ames Research Center in California and from there he came back East as the Director of the Goddard Space Flight Center. After leaving NASA, Tom joined the Martin Marietta Corporation 1982 as Vice President of Aerospace Research and Engineering. Later he became President and Chief Operating Officer of the Martin Marietta Corporation and retired as Executive Vice President of Lockheed Martin Corporation. He has been on many boards of large corporations and has led several studies in support of congress and federal agencies regarding various NASA and space related activities. He is also a willing speaker and continues to be involved with NASA and other federal agencies in support of the space program.

Israel "Is" Taback

"Is" was the Deputy Project Manager and Chief Technical Manager for LO and Viking. He retired from NASA shortly after Viking and joined the Bionetics Corporation. He continued to work with NASA for many years as an aerospace consultant on many technical problems until his death in August of 2008. He received many awards during his career and was also awarded an honorary Ph. D in science from Old Dominion University. We have all kept and used the many lessons we learned from him.

Bill Boyer

Bill was Missions Operations Manager for Lunar Orbiter. Before Lunar Orbiter, Bill played a major role in setting up the worldwide tracking network for Apollo. Even before that, in the early days of radar, he used that capability to position aircraft during the atomic bomb testing on Bikini Atoll. During Viking, he served as Deputy for Mission Operations. After Viking, he moved into the Large Space Structures Program in which NASA was experimenting with building large
structures in space, and continued to contribute to this program until his retirement from NASA in 1985.

Otis Childress

During Lunar Orbiter, Otis was assigned to many areas, including the on-board pyrotechnics and spacecraft integration. Then he was placed in charge of getting a clean room built at the Cape for the LO vehicle. After launch, Otis headed to California to take his place on the flight operations team. On Viking he held many jobs, including the Orbiter Manager, which he held until leaving Viking. After Viking, Otis was assigned as Head of the Projects
Integration Office at Langley, then as Chief Engineer for the Rotor Systems Research Aircraft, which studied innovative systems on helicopters. He then served as the Project Manager for the National Rotorcraft Noise Reduction Project that won the Howard Hughes Award.

Norman "Norm" Crabill

After Viking, Norm initiated the Storm Hazards Program, and then after retirement he was responsible for the development of the technique for lightning-proofing composite aircraft and the system for delivering weather data to the cockpit using satellite downlink. These systems are in general use today throughout the industry.

John Graham

Prior to Lunar Orbiter, John had been the head of recovery operations for the Mercury Project. In fact, the whole family got involved, as they used John's father's Chesapeake Bay crabbing boat for a large portion of the experimental recovery operations in the early days of the project prior to the time the whole Apollo Program moved to Houston. As the Lunar Orbiter Project developed, John became the manager responsible for developing the ground-based photographic processing system housed at the tracking stations and in the JPL Space Flight Operations Facility (SFOF). During the missions, he managed the photo processing center at the SFOF. After the Lunar Orbiter Project, John went to Houston, but as the Viking Project solidified, he came back to Langley and took the role of the Head of Viking Mission Operations. Prior to Viking operations, John retired from NASA to take over the family business at the unexpected death of his father.

Matthew (Matt) Grogan

Matt joined The Boeing Company in 1960 and became the Lead Orbit Determination Engineer for the Lunar Orbiter Project in 1965. He served in this capacity for all five Lunar Orbiter missions. After Lunar Orbiter, he joined the Manned Spaceflight Center in Houston and served as a member of the Apollo navigation team for missions 8 through 15, and helped develop the precision landing techniques used on Apollo missions 12 through 17. He and the Apollo Missions Operations Team received the Presidential Medal of Freedom for their efforts on Apollo 13. After Apollo, Matt accepted a position at Martin Marietta, where he served as Sequence Development Chief for the Viking Flight Team. After Viking, he was involved in several technical and management assignments for Martin Marietta, including Systems Engineering Director for the USAF Peacekeeper ICBM development. He retired in 1988 and continues to be involved in multiple community support activities in his hometown of Littleton, CO.

Angelo (Gus) Guastaferro

After Viking, Gus became the Program Manager for NASA's Rotor Systems Research Aircraft Project at Langley. Then, on to Washington, DC to become the Director of Planetary Programs in NASA's Office of Space Science. He later moved west to become the Deputy Director of NASA's Ames Research Center. Gus retired from NASA to accept the position of Vice President, Lockheed Missiles and Space for Civil Space Programs. He now works with the Dean of the William and Mary Mason School of Business as an advisor in the school's MBA program, and he teaches in the Florida Institute of Technology Masters Program. He continues to sit on many boards and to consult with NASA.

Gentry Lee

Before relating where Gentry is today, I must tell you a little story about him. On one of our many meetings at Martin Marietta in Denver, several of us went to Gentry's home for drinks and then to dinner.

During the happy hour, Gentry mentioned that his back door lock wasn't working properly. This information was given to several engineers who had had a drink or two, and we could not pass up the challenge.

We immediately gathered all the tools that were nearby, including screw drivers, pliers, hammers, chain saws, and so on, and began to work on the lock. We had that door completely apart, diagnosed, corrected, and reassembled in just a few minutes. Gentry caught hell the rest of the evening because "purely theoretical types like him just couldn't do anything useful. Only engineers could make things happen."

What did Gentry do after Viking?

Well, he produced the Cosmos Series with Carl Sagan. He wrote a four-volume science fiction series entitled Rama with Arthur C. Clarke. All four books were New York Times Best Sellers and were translated in over twenty-five languages. Since his collaboration with Mr. Clarke, he has written three more highly successful solo novels. He now serves as the Chief Engineer for the Solar System Exploration Directorate at the Jet Propulsion Laboratory in Pasadena, California.

What are the rest of us engineers doing?

Well, maybe just fixing locks.

Burt Lightner

Burt was responsible for developing the launch windows and periods for the five LO flights and both of the Viking launches, considering the tracking constraints, conditions at arrival at the Moon or Mars, and launch vehicle constraints. After the Viking launches, he headed west to take his position as part of the flight operations team in the mission planning area. After Viking, Burt served as the Project Manager for the ACCESS Experiment on the Space Shuttle to demonstrate astronaut EVA assembly of an erectable space structure. Subsequently, he was Project Manager for the Shuttle retrieval and initial ground processing of NASA's Long Duration Exposure Facility, which had been in space for several years in order to understand the effects of the space environment on different materials. Burt retired from NASA in 1995.

Uriel (Woody) Lovelace

After Viking, Woody spent some time at Langley developing experiments using large space structures. Then, he went on to NASA Headquarters to study various options for returning humans to the Moon and Mars. Finally, back at Langley, he was liaison with Houston for the International Space Station efforts until his retirement.

William (Bill) Michael

During Lunar Orbiter, Bill had been the leader of the Radio Science Team and Principal Investigator for the Lunar Geodesy Experiment. After leaving Viking as head of the Radio Science Team, Bill spent time as Special Assistant to the Langley Director, and then served for 6 years in the National Academy of Science. He also served as an Associate Director for the Jefferson Lab in Newport News VA and has traveled extensively giving papers on the understanding of the gravitational fields and structure of the Moon and Mars.

John Newcomb

After Viking, John headed a group of project managers involved in a wide spectrum of projects, from placing experiments on the Space Shuttle to putting pylons on fighter aircraft. He headed NASA's Physics and Chemistry Experiments in Space Program, which was to become the core of the newly formed NASA Microgravity Program. This program conducted experiments in the apparent gravity-free environment of space that could not be conducted here on Earth.

After retiring from NASA in 1984 he began consulting with NASA, taught courses on project management and system engineering, and ran NASA's Advanced Project Management course, training project managers from the different centers and from other government agencies. He continues to consult with NASA, teach courses, and write a book.

Dale Shellhorn

After Lunar Orbiter, Dale spent 3 years in Houston supporting the Apollo Program. He served as Chief Engineer for the Personal Rapid Transit System Program at University of West Virginia, which produced 70 all-electric unmanned vehicles connecting campuses and dorms for the student population. The rest of his Boeing career was spent modifying AF B-52 Bombers and as Chief Engineer and Program Manager for the B-1B Bombers. Finally, he became the Program Manager and VP for the Northrop-Grumman B-2 Program. He now enjoys Tucson and the warm climate that is conducive to his game of golf.

Paul Siemers

Paul came back to Langley after completing graduate work for his Master's Degree just in time to be the staff man for the Lunar Orbiter (LO) Mission Director.
After LO, Paul signed on to Viking to serve as the Aerothermodynamics Engineer responsible for the testing, analyses, and development of databases required for the design of the entry and descent subsystems, which consisted of the Entry Vehicle aeroshell, base cover, and the parachute. After Viking, he conducted entry-related experiments on the Space Shuttle and supported the entry performance analysis of the Orbiter. The results of his research on entry vehicle aerodynamics, aerothermodynamics, and parachutes are major texts in those fields today.

Gerald Soffen

Gerry was the chief scientist for Viking and headed the Viking Science Team. After Viking, Gerry went to NASA Headquarters as Director of Life Sciences, then on to Goddard Spaceflight Center, where he was the Director of University Programs. Although he was disappointed about the ambiguous findings from Viking, he never gave up his passion for finding extraterrestrial life.

Bob Tolson

During Lunar Orbiter, Bob had been a member of the Radio Science Team. Bob's Initial position with Viking was as Navigation Manager, responsible for navigation from Centaur burnout to landing. He later joined the Viking Radio Science Team, and after the primary mission, he initiated the Phobos-Demos Encounter Experiment in which the Viking Orbiter flew close to both Phobos and Demos, thereby allowing the determination of the mass of the Martian satellites and pictures with 1-meter resolution. Later, he served in many capacities at Langley, including Chief Scientist. He subsequently served as Professor at George Washington University, University of Maryland, and North Carolina State University, and he is currently a co-investigator on the MAVEN mission.

Donald (Don) Ward

During Lunar Orbiter, Don coordinated activities at the Cape for all of the five launches. He was the Cape Facilitator for both of the Viking launches. After the launches, he came home and made one final trip to California to be there with the rest of the team for the landings. After the Viking launch, Don became part of the management team for the Advanced Technology Laboratory (ATL) to be placed in the Space Shuttle that was being developed at that time. He then was the construction manager for the National Transonic Facility (NTF), which is a large tunnel that uses nitrogen at very cold temperatures and can test models at and near the speed of sound.

References

Publications

1. American Geophysical Union; *Scientific Results of the Viking Project*; Washington D. C. ; ISBN0--87590-207- 3; Library of Congress Catalogue Card Number, 77-- 23719

2. Biemann, Hans---Peter; *The Vikings of '76*; Published by Hans---Peter Biemann, 175 Brattle Street, Cambridge, MA, 02138, USA; 1977; Library of Congress Catalogue No. 77-71945

3. Byers, Bruce; *Destination Moon, A History of the Lunar Orbiter Program*; NASA, History Office, Washington D. C.; NASA TM X--3487; April 1997

4. Crabill, Norman; *Minutes of the Landing Site Selection Team Meetings, number 1 through* 48, dating from 17 June, 1976 through August 30, 1976

5. Digregorio, Barry; Levin, Dr. Gilbert; Stratt, Patricia Ann; *Mars The Living Planet*; Frog, Ltd.; 1997

6. Ezell, Clinton Edward, Ezell, Linda Neuman; *On Mars, Exploration of the Red Planet*, 1958, 1978; NASA SP 4212; 1984

7. French, Bevan, M.; Mars, *The Viking Discoveries*; Office of Space Sciences, NASA; EP---146; October 1977.

8. Hansen, James; *Spaceflight Revolution; NASA History Series*, NASA SP-4308; NASA Washington D. C.; 1995

9. Lightner, E. B.; *LD/ED Strategy and Daily Launch*

Windows; Memo to Distribution; June 16, 1975.

10. Martin Marietta Corporation, Denver Division; *Viking Lander, "As Built" Performance Capabilities*; NAS1-9000; June 1976

11. Masursky, Harold; Crabill, Norman ; *Viking Site Selection and Certification*; Prepared by Langley research Center; NASA Scientific Technical and Information Branch; Washington, DC; 1981

12. NASA; *Viking 1 – Early Results, SP-408*; August 1976.

13. Reiber, Duke, B, Editor; *The Mars Conference; Volume 71, Science and Technology Series, Volume 71*;American Astronautical Society; 1988

14. Reichardt, Tony; Canaveral Junior; web address: http://www.airspacemag.com/multimedia/canavera l-junior-6283414/

15. Shortall, Joseph, A; *History of Wallops Station*; NASA, Wallops Station, Wallops Island VA, 23337; Under contract NAS--6--1657; 1971

16. The Boeing Company, Space Division; *The Lunar Orbiter*; Seattle, Washington; April 1966

17. Ward, Donald; *One Engineer's Life Relived, An Autobiography*; Unpublished

Interviews, Notes and Correspondence

18. Boyer, William – 2010

19. Bruton, Dempsey – 2013 – 2014

20. Childress, Otis – 2013 – 2014

21. Crabill, Norman – 2013 -2015

22. Cuddihy, Fillmore – 2012 – 2014

23. Digregorio, Barry – 2013 - 2015

24. Graham, John – 2012 – 2014

25. Grogan, Matthew – 2012 – 2014

26. Levin, PhD, Gilbert - 2012 – 2015

27. Lightner, Burt – 2012 – 2014

28. Lovelace, Uriel – 2013 – 2015

29. Seamers, Paul – 2012 – 2014

30. Shellhorn, Dale – 2013 – 2015

31. Tillman, Rachel – 2014 - 2015

32. Tolson, PhD Robert – 2013 – 2014

Website utilized for reference

33. http://nssdc.gsfc.nasa.gov

END NOTES

Throughout this book, the author refers to information cited from the references on the previous page. The nomenclature is as follows:

Reference: refers to the number identifying each reference on the previous page.

Page: refers to the specific page(s) within the reference.

For example, the first reference lists Reference 8: Page 315.

This refers to Hansen, James; Spaceflight Revolution; NASA History Series, NASA SP-4308; NASA Washington D. C.; 1995 And

specifically page 315.

Chapter	Note	Reference
A Bunch of Plumbers	1	Reference 8: Page 315
Chapter One	1	Reference 33
	2	Reference 2: Page 39
Chapter Three	1	Reference 14
Chapter Five	1	Reference 21
	2	Reference 19
Chapter Six	1	Reference 21
Chapter Ten	1	Reference 8: Page 221
	2	Reference 33
	3	Reference 33

Chapter	Note	Reference
	4	Reference 33
	5	Reference 33
Chapter Eleven	1	Reference 3: Page 40
	2	Reference 32
	3	Reference 8: Page 321
Chapter Twelve	1	Reference 3 – Page 40
	2	Reference 8: Page 326, Reference 3: Page 56
	3	Reference 3: Page 95
Chapter Fourteen	1	Reference 3: Page 70 & 83
	2	Reference 8: Page 330
Chapter Nineteen	1	Reference 3: Page 228
Chapter Twenty-Two	1	Reference 18
Chapter Twenty-Five	1	Reference 20
	2	Reference 8: page 345
	3	Reference 24
	4	Reference 8: Page 345
Chapter Thirty-Two	1	Reference 3: Page 256

Chapter	Note	Reference
Chapter Thirty-Four	1	Reference 21
Chapter Thirty-Nine	1	Reference 25
	2	Reference 32
	3	Reference 25
Chapter Forty	1	Reference 6: Page 151
	2	Reference 6: Page 151
Chapter Forty-One	1	Reference 6: Page 216
Chapter Forty-Two	1	Reference 6: Page 216
Chapter Forty-Four	1	Reference 6: Page 190
Chapter Forty-Five	1	Reference 6: Page 278
	2	Reference 6: Page 314
	3	Reference 32
	4	Reference 6: Page 289
	5	Reference 6: Page 291
	6	Reference 33
Chapter Forty-Six	1	Reference 6:Page 302
	2	Reference 6: Page 313

Chapter	Note	Reference
Chapter Fifty	1	Reference 29
Chapter Fifty-Two	1	Reference 33
Chapter Fifty-Three	1	Reference 6: Page 55
Chapter Fifty-Five	1	Reference 9 & 27
	2	Reference 9
	3	Reference 27
	4	Reference 17: Page 173
	5	Reference 17 Page 112 — Reference 9
	6	Reference 17: Page 112 — Reference 9
	7	Reference 17: Page 113
	8	Reference 17: Page 112
	9	Reference 20
	10	Reference 17: Page 113
Chapter Fifty-Seven	1	Reference 6: Page 331
	2	Reference 4: LSS Meeting No. 6
Chapter Sixty-Three	1	Reference 4
	2	Reference 4
	3	Reference 4

Chapter	Note	Reference
Chapter Sixty-Four	1	Reference 6: Page 355
	2	Reference 2: Page 135
Chapter Sixty-Six	1	Reference 12: Page 41
	2	Reference 6: Page 417
	3	Reference 6: Page 390 also Reference 31
	4	Reference 7: Page 17
	5	Reference 6: Page 392
	6	Reference 12: Page 51
	7	Reference 6: Page 417
	8	Reference 7: Page 14
	9	Reference 7: Page 14
	10	Reference 7: Page 17
	11	Reference 7: Page 26
Chapter Sixty-Seven	1	Reference 7: Page 20
	2	Reference 6: Pages 404-405
	3	Reference 6: Pages 404-405
	4	Reference 6: Page 405-406
	5	Reference 6: Page 406

Chapter	Note	Reference
	6	Reference 6: Page 406
	7	Reference 26
	8	Reference 26
	9	Reference 6: Page 407
	10	Reference 6: Page 408
	11	Reference 6: Page 409
	12	Reference 26
	13	Reference 26
	14	Reference 26
	15	Reference 5: Page 110
	16	Reference 5: Page 110
	17	Reference 5: Page 118
	18	Reference 5: Page 120
	19	Reference 5: Page 125
Chapter Sixty-Eight	1	Reference 13: Page 182
	2	Reference 13: Page 183
	3	Reference 13: Page 183
	4	Reference 13: Page 206

Chapter	Note	Reference
	5	Reference 13: Page 206
Chapter Sixty-Nine	1	Reference 26
	2	Reference 26
	3	Reference 23
	4	Reference 26

APPENDICES

Appendix A –Lunar Orbiter Mission Advisors

Credit NASA: JPL-P 7328

NASA LANGLEY MISSION ADVISORS

TOP ROW

W. R. Sanborn, L. P. Daspit, U. K. Lovelace, M. J. Pilny, G. W. Brewer, N. A. Holmberg, L. J. Kosofsky, R. W. Mixon, J. C. Moorman, R. H. Sproull, Bronson (EK)

MIDDLE ROW

P. M. Siemers, T.W.E. Hankison, L. A. Wallace, G. C. Broome, N. L. Crabill, H. L. Smith, J. R. Unangst, J. J. Lyons (TBC), W. T. Bundick, P. Froome (EK), E. A. Brummer, K. L. Wadlin

BOTTOM ROW

R. R. Peterson, J. A. Wolicki, G. Schaber, D. J. Carter, J. F. Newcomb, J. M. Drozdowski, W. J. Boyer, J. D. Timmons, J. E. Harris, C. H. Nelson, I. Taback, D. D. Webb

Appendix B-Lunar Orbiter Red Operations Team

Credit: NASA-JPL

RED TEAM

TOP ROW
A. J. Khatib (JPL), R. Wallace (JPL), W. Sjogren (JPL),
J. P. Brenkle (JPL), R. H. Smith

SECOND ROW
R. R. Kaufman, H. J. Afker, G. D. Barrow, L. A. Schaut,
J. G. Hoos, D. C. Shellhorn, M. M. Grogan, P. Huntress,
L. Neilson

THIRD ROW
O. G. Fitz, W. G. Moyer, M. D. Mills, W. D. Allingham,
R. H. Daly, T. R. Cox, R. N. Hahn, R. R. Paulson

FOURTH ROW
R. L. Farmer, W. H. Anderst, J. E. Bandall, J. F. Smith,
G. K. Stewart, H. R. Newkirk, H. B. Woody, D. E. Johnson,
D. B. Bunger

BOTTOM ROW
E. J. Hamilton, J. L. Griggs, R. L. McCauley, R. F. Caruso,
C. A. Eagan, K. Babunes, J. J. Brossard, J. C. Graves,
L. J. DeRyder (LRC), L. M. Brooks, R. G. Erwood

Appendix C –Lunar Orbiter White Operations Team

Credit: NASA-JPL-P 7329

WHITE TEAM

TOP ROW:
W. A. Swenning, F. M. DeGroot, A. N. Rumpf, R. P. RUDD,
C. S. Gooskin, C. W. Johnson, D. R. Shelmandine, P. E. Hong,
S. R. Zimmerman

SECOND ROW:
R. R. Sexauer, G. R. Schmeer, W. E. Coon, J. R. Malcolm,
D. M. Ownbey, B. J. Slind

THIRD ROW:
J. I. Nadeau, J. R. Williams, S. D. Bradshaw, F. W. Locke,
S. C. Brink, D. S. McKellar, B. A. Paquette, J. A. Kinahill

BOTTOM ROW:
K. L. Peterson, W. B. Hemphill, H. H. O'Daniels, D. M. Choi,
G. G. Grody, G. J. Burmeister, G. M. Fuller, R. R. Peterson (LRC),
G. W. Hettrick

Appendix D –Lunar Orbiter Blue Operations Team

Credit: NASA-JPL-P 7321

BLUE TEAM

TOP ROW

J. F. Gilbert, J. D. Kyers, J. A. Beacon, R. E. Stevenson,
W. R. Sanborn (LRC), D. B. Barrs, P. M. Blood, R. F. Johnson,
K. D. Hill, D. Vanzandt

SECOND ROW

A. Caticchio, G. K. Essmier, D. H. Campbell, M. C. Wilkinson,
G. R. Knapp, G. A. Price, M. V. Farris, J. G. Gray, D. N. MacLean,
R. E. Hansen, W. P. Marsh

THIRD ROW

F. Leppla, D. F. Gallagher, A. W. Snow, C. M. Poe, P. A. Pierson,
V. Kennicott, V. C. Benge, D. J. Fries, G. R. Caldwell, T. A. Grant,
W. L. Bunce, T. J. Hansen

BOTTOM ROW

L. M. Brooks, R. Miller (JPL), R. C. Rondeau, R. L. Balesh,
T. R. Steury, W. R. Burr, J. H. Stachurski, R. V. Johnson,
B. Wittman (EK), R. W. Delteare, J. Olson, B. M. O'Brien,
G. L. Clark

Appendix E – Viking Science Teams

Gerald Soffen - Viking Chief Scientist
Conway Snyder – Viking Orbiter Science Group Chief
Richard. S. Young – NASA Program Scientist
G. Calvin Broome – Viking Lander Science Group Chief

is   *Molecular Analysis*

Orbiter Imaging

Michael H. Carr, USGS, Menlo Park
William A. Baum, Lowell Observatory
Karl R. Blasius, Science Applications
Geoffrey Briggs, JPL
James A. Cutts, Science Applications
Thomas C. Duxbury, JPL
Ronald Greeley, University of Santa Clara
John E. Guest, University of London, England
Keith A. Howard, USGS, Reston
Harold Masursky, USGS, Flagstaff
Bradford A. Smith, University of Arizona
Lawrence A. Soderblom, USGS, Flagstaff
John B. Wellman, JPL
Joseph Veverka, Cornell University

Lander Imaging

Thomas A. Mutch, Brown University
Alan B. Binder, Science Applications
Friedrich O. Huck, LRC
Elliott C. Levinthal, Stanford University
Sidney Liebes, Jr., Stanford University
Elliott C. Morris, USGS
James A. Pollack, Ames Research Center
Carl Sagan, Cornell University

Biology

Harold P. Klein, Ames Research Center
Norman H. Horowitz, Caltech
Joshua Lederberg, Stanford University
Gilbert V. Levin, Biospherics
Vance I. Oyama, Ames Research Center
Alexander Rich, MIT

Thermal Mapping

Hugh H. Kieffer, UCLA
Stillman C. Chase, Santa Barbara Research Center
Ellis D. Miner, JPL
Guido Munch, Caltech
Gerry Neugebauer, Caltech

Water Vapor Mapping

C. B. Farmer, JPL
Donald W. Davies, JPL
Dan LaPorte, Santa Barbara Research Center

Entry Science

Alfred O. C. Nier, University of Minnesota
William B. Hanson, University of Texas
Michael B. McElroy, Harvard University
Alfred Seiff, Ames Research Center
Nelson W. Spencer, Goddard Space Flight Center

Klaus Biemann, MIT
Duwayne M. Anderson, Cold Regions Research and Engineering Laboratory, U.S. Army
Alfred O. C. Nier, University of Minnesota
Leslie E. Orgel, Salk Institute
John Oro, University of Houston
Tobias Owen, State University of New York
Priestley Toulmin III, USGS, Reston
Harold C. Urey, University of California, San Diego

Inorganic Chemical

Priestley Toulmin III, USGS, Reston
Alex K. Baird, Pomona College
Benton C. Clark, Martin Marietta Aerospace
Klaus Keil, University of New Mexico
Harry J. Rose, USGS, Reston

Meteorology

Seymour L. Hess, Florida State University
Robert M. Henry, LRC
Conway B. Leovy, University of Washington
Jack A. Ryan, California State University, Fullerton
James E. Tillman, University of Washington

Seismology

Don L. Anderson, Caltech
Fred Duennebier, University of Texas
Robert L. Kovach, Stanford University
Gary V. Latham, University of Texas
George Sutton, University of Hawaii
Nafi Toksöz, MIT

Physical Properties

Richard Shorthill, University of Utah
Robert E. Hutton, TRW
Henry J. Moore II, USGS, Menlo Park
Ronald F. Scott, Caltech

Magnetic Properties

Robert B. Hargraves, Princeton University

Radio Science

William H. Michael, LRC
George Born, JPL
Joseph P. Brenkle, JPL
Dan L. Cain, JPL
J. G. Davies, University of Manchester, England
Gunnar Fjeldbo, JPL
Mario D. Grossi, Raytheon
Robert Reasenberg, MIT
Irwin I. Shapiro, MIT
Charles T. Stelzried, JPL
Robert H. Tolson, LRC
G. Leonard Tyler, Stanford University

Credit: Reference 1, Page 3961

Appendix F Viking Landing Site
Selection Team

The Landing Site Selection Team, in reality, was comprised of any and all individuals who could offer some advice or support relevant to finding the most suitable landing sites on Mars. The leaders, staff and official list of members are shown below. However, this was, in truth, a huge team effort, conducted by all members of the different science teams that were analyzing and interpreting the incoming data in reference to their areas of expertise, and then integrating all of these varied viewpoints in order to arrive at a consensus. In like manner, the flight operations team continued to develop multiple mission options and rendered engineering assessments as the possible sites were considered.

Leaders

Thomas Young, Mission Director: NASA Langley Research Center

Gentry Lee, Science Analysis and Mission Planning Director: Martin Marietta Corporation

Harold Masursky, Leader: U S Geological Survey

Norman Crabill, Executive Secretary: NASA Langley Research Center

---

Staff

Doyle Vogt: - Martin Marietta Corporation
John Newcomb: - NASA Langley Research Center

Appendix F Viking Landing Site Selection Team Members

William Baum, Lowell Observatory
Geoffrey Briggs, Jet Propulsion Laboratory
Michael Carr, U S Geological Survey
George Colton, U S Geological Survey
C. Barney Farmer, Jet Propulsion Laboratory
Noel Hinners, NASA Headquarters
Carol Hodges, U S Geological Survey
Hugh Kieffer, University of California at Los Angeles
Terrence Kreidler, U S Geological Survey
Joshua Lederberg, Stanford University
Conway Leovy, University of Washington
John MaCauley, U S Geological Survey
Charles Meyer, U S Geological Survey
Daniel Milton, U S Geological Survey
Henry Moore, U S Geological Survey
Elliot Morris, U S Geological Survey
ThomasMutch, Brown University
Kenneth Murray, U S Geological Survey
Joseph O'Connor, U S Geological Survey
Tobias Owen State, University of New York
James Pettengill, U S Geological Survey
Richard Pike, U S Geological Survey
James Porter, Martin Marietta Corporation
Carl Sagan, Cornell University
Stephen Saunders, Jet Propulsion Laboratory
Daniel Scott, U S Geological Survey
Bradford Smith, University of Arizona
Laurence Soderblom, U S Geological Survey
Gerald Soffen, NASA Langley Research Center
Desiree Stuart-Alexander, U S Geological Survey
Newell Trask, U S Geological Survey
Alta Walker, U S Geological Survey
Mareta West, U S Geological Survey
Donald Wilhelms, U S Geological Survey

Reference 12: Page 67

Appendix G Viking Team Key Personnel

NASA Headquarters

N. W. Hinners, Associate Administrator for Space Science
R. S. Kraemer, Director, Planetary Programs
W. Jakobowski, Program Manager
R. S. Young, Program Scientist

Langley Research Center

E. Cortright, Director – May 1968 – September 1975
D. P. Hearth, Director – September 1975 – November 1984

Viking Flight Team

J. S. Martin, Jr. , Project Manager
T. Young, Mission Director
G. A. Soffen, Project Scientist
J. D. Goodlette, Chief Engineer
G. Lee, Science Analysis and Mission Planning Director
P. T. Lyman, Spacecraft Performance and Flight Path Analysis Director
M. J. Alazard, Mission Control Director
N. Gianopulos, Mission Control Computing Center Systems Engineer
E. Van Ness, Senior Staff (External Affairs)
R. L. Crabtree, Deputy Mission Director
L. Kingsland, Deputy Mission Director (Planning)
W. Snyder, Orbiter Science Group Chief
G. C. Broome, Lander Science Group Chief
Porter, Mission Planning Group

R. A. Ploscaj, Orbiter Performance Analysis

R. W. Sjostrom, Lander Performance Analysis

W. J. O'Neil, Flight Path Analysis

M. M. Grogan, Sequence Development

L. S. Canin, Flight Control

D. D. Gordon, Data Support

W. B. Green, Image Processing Staff Leader

H. Masursky, Landing Site Selection Team Leader

N. Crabill, Landing Site Selection Team Secretary

D. Vogt, Landing Site Selection Team Staff

J. Newcomb, Landing Site Selection Team Staff

K. S. Watkins, Administrative Support Office

K. W. Graham, Ground Data Systems Support R.

J. Polutchko, Lander Support Office Chief

K. H. Farley, Lander Support Engineering

F. D. Nold, Lander Support Operations

B. A. Claussen, Lander Support Software

ABOUT THE AUTHOR

John had many adventures while attending Virginia Tech, and alternately working in the NASA Langley Research Center's wind tunnels, laboratories, and at Langley's rocket launching site, Wallops Island. After graduating from Virginia Tech John joined the NASA Langley Research Center as a young engineer and quickly became involved in two of NASA's early space missions. The first was the Lunar Orbiter Project which placed 5 Lunar Orbiters around the Moon and photographed the Apollo landing sites. The Second was the Viking Project which placed two landers on the surface of Mars, in 1976. In the book, *A Bunch of Plumbers*, John gives you an exclusive behind the scenes look at these two extraordinary missions. This is a history of those missions, through John's eyes, told with stories, events, and human interest.

After Viking, John headed a group of project managers performing projects as varied as putting experiments on the NASA Space Shuttle to putting pylons on fighter aircraft.

He then headed NASA's Physics and Chemistry Experiments in Space Program which performed experiments in the free fall environment of the NASA's

Space Shuttle and The International Space Station. This program, which involved many NASA centers and experimenters from all over the country, became the core of NASA's Microgravity Program.

After retiring 1984, he consulted with NASA, headed an office in Washington DC which supported the Microgravity Program and taught courses on project management and system engineering. He also ran NASA's Advanced Project Management course, training project managers from the different centers and from other government agencies. He continues to consult with NASA, and teach courses on project management and systems engineering.

Throughout his career John has authored or co-authored over 35 technical papers varying in subject matter from interplanetary trajectory design and guidance theory to Total Quality Management.

Along with his professional endeavors he has found time to indulge his hobbies such as sailing and skiing. He teamed with a fellow sailor, Harry Sindle, and came in second in the 1964 Olympic trials, sailing in the Flying Dutchman class. He then won the US National Championships in 1965 and competed in the Flying Dutchman World Championships in Alassio, Italy. Now his passion is skiing and he has skied various locations on the East Coast, up and down the Rockies, and dozens of different places in Europe.

He lives in Gloucester, Virginia with his wife, Peggy and granddaughter, Alexus.

Made in the USA
Columbia, SC
16 July 2017